Studies in History and Philosophy of Science

Volume 63

Studies in History and Philosophy of Science is a peer-reviewed book series, dedicated to the history of science and historically informed philosophy of science. The series publishes original scholarship in various related areas, including new directions in epistemology and the history of knowledge within global and colonial contexts. It includes monographs, edited collections, and translations of primary sources in the English language. These cover a broad temporal spectrum, from antiquity to modernity, and all regions of the world.

Alberto Fragio • Josefa Ros Velasco
Martina Philippi • Cornelius Borck
Editors

Hans Blumenberg's History and Philosophy of Science

Springer

Editors
Alberto Fragio
Departamento de Humanidades, Torre III
Universidad Autónoma Metropolitana -
Unidad Cuajimalpa
Mexico City, Mexico

Josefa Ros Velasco
Department of Humanities
ESIC University
Madrid, Spain

Martina Philippi
Kulturwissenschaften | Fach Philosophie
Universität Paderborn
Paderborn, Nordrhein-Westfalen, Germany

Cornelius Borck
Institut für Medizingeschichte und
Wissenschaftsforschung
Universität zu Lübeck
Lübeck, Schleswig-Holstein, Germany

ISSN 1871-7381 ISSN 2215-1958 (electronic)
Studies in History and Philosophy of Science
ISBN 978-3-031-78250-3 ISBN 978-3-031-78251-0 (eBook)
https://doi.org/10.1007/978-3-031-78251-0

This Springer imprint is published by the registered company Springer Nature Switzerland AG
The registered company address is: Gewerbestrasse 11, 6330 Cham, Switzerland

Contents

The original version of the book has been revised. A correction to this book can be found at https://doi.org/10.1007/978-3-031-78251-0_10

Part I
A Philosophical History of Sciences

An Introduction to Hans Blumenberg's History and Philosophy of Science

Alberto Fragio and Josefa Ros Velasco

Abstract A prominent German philosopher in the post-WW II era, Hans Blumenberg (1920–1996) skilfully connected the continental tradition of philosophy with conceptual history and metaphorology. Some of his major works, such as *Paradigms for a Metaphorology*, *The Legitimacy of the Modern Age*, *The Genesis of the Copernican World* or *Work on Myth* have already been translated into English. And a worldwide rising interest in his philosophy can be registered today. Comparably less attention has been paid though by the scholarly literature to Blumenberg's way of doing history of science and its impact on current historians and philosophers of science. In order to fill this gap, this volume aims to explore the methodological instruments Blumenberg developed, and how we may use them to do history and philosophy of science. The main goal motivating our endeavour is to shed some light on a somehow neglected tradition, which found little room within the dominant analytical philosophy of science. What is at stake is some sort of non-received view in contemporary history and philosophy of science, which may be termed 'continental history and philosophy of science.' Standard references on this topic include the now classic book edited by Gary Gutting, *Continental Philosophy of Science* (2005) and, in a similar vein, Mary Domski and Michael Dickson's

---

This chapter has been written within the framework of the research projects PID2021-128606NB100 and PID2022-141667NB-I00 funded by the Spanish Ministry of Science and Innovation (Spanish Government), and of the European Union's Horizon 2020 research and innovation programme under the Marie Skłodowska-Curie grant agreement No 847635. The remarks and criticisms of the research group HIST-EX helped improve the final version of the manuscript. We would also like to thank Tessa Marzotto Caotorta for the English revision. Any remaining mistakes or flaws are, of course, our sole responsibility.

A. Fragio (✉)
Universidad Autónoma Metropolitana, Unidad Cuajimalpa, Mexico City, Mexico
e-mail: afragio@cua.uam.mx

J. Ros Velasco
Department of Humanities, ESIC University, Madrid, Spain
e-mail: josros@ucm.es

A. Fragio et al. (eds.), *Hans Blumenberg's History and Philosophy of Science*, Studies in History and Philosophy of Science 63,
https://doi.org/10.1007/978-3-031-78251-0_1

Discourse on a New Method: Reinvigorating the Marriage of History and Philosophy of Science (2010). Our book will serve to map Blumenberg's contribution to the continental history and philosophy of science. This will not only add new input to the field of history and philosophy of science, but also opens a new area in the scholarship on Blumenberg's works.

1 What Is Continental History and Philosophy of Science?[1]

Gary Gutting (2005), a former professor at Notre Dame University, provides some insights on what continental philosophy of science is. In Gutting's opinion, the very beginning of philosophy of science as an autonomous knowledge can be found in Immanuel Kant's *Critique of Pure Reason* (published in 1781). Kant's approach to Galilei's, Descartes', and Newton's contributions to modern science, and their standing out compared to traditional philosophy, can be seen as involving a deep shift in the understanding of the *a priori* conditions. As is well known, empirical methodologies introduced during the seventeenth century allowed to separate philosophical knowledge from scientific knowledge. In this respect, Gutting suggests "a rough but useful" (2005, p. 1) taxonomy of continental philosophy of science based on different attitudes regarding the relationship between philosophy and science. The empirical or positivist attitude sees science as autonomous, and philosophy as a secondary reflection which explicits the outcomes achieved by science. The second attitude, closely related to Kantian criticism, sees science as offering original knowledge, but philosophy as evidencing the conditions that make scientific knowledge possible. In other words, while philosophical justification is supposed to depend on the validity of scientific knowledge, philosophy is supposed to provide transcendental truths. The third and last attitude, the metaphysical one, believes that the access to philosophical truths is independent from and even superior to science. Furthermore, according to Gutting's taxonomy, the empirical attitude is common among scientists and philosophers committed to science, such as Mach, Poincaré, and Duhem. The second attitude, the critical one, is championed by French and German neo-Kantianism, in particular by authors such as Brunschwicg or Bachelard, for the French neo-Kantianism, and by Cohen, Natorp, and Cassirer, for the German neo-Kantianism. And the metaphysical attitude would be typical of authors such as Bergson, Dilthey, and Husserl, among others.

Along similar lines, one might attempt to also map the continental history of science. In this sense, a volume on Blumenberg's contributions to history and philosophy of science provides a good opportunity to resume such an endeavour. However, it will soon become clear that Blumenberg's understanding of the history and philosophy of science does not fit Gutting's taxonomy. While being close to the Kantian attitude, Blumenberg develops a personal take on philosophy of science, which goes

[1] This section is partially based on Fragio Gistau (2020).

beyond the idea that philosophy should either be a strict science or a tool for exploring the conditions of possibility of scientific knowledge.

2 The Ideal of a Scientific Philosophy

In order to provide some philosophical background for Blumenberg's views on the history and philosophy of science, it is worth adding some remarks on the ideal of scientificity in philosophy. Blumenberg's metaphorology is in part a revision of phenomenology, but it has lost one of Husserl's original impulses: to turn philosophy into strict science. Nowadays it may seem naïve to try to turn philosophy into a science, but this was indeed the purpose of Husserl's transcendental phenomenology. This was not at all the aim of Blumenberg's reception of phenomenology, yet his metaphorology is placed in the aftermath of one of the biggest projects to turn philosophy into a strict science by means of a deep epistemological restructuring.

In a well-known series of articles and books, the American philosopher Michael Friedman (2001, 2012) suggests that the ideal of a scientific philosophy first emerged in the mid-nineteenth century as an adverse reaction to post-Kantian German idealism and its metaphysical speculative character (see Engler and Renn, 2018; Morris, 2020). Hermann von Helmholtz voiced a climate of hostility and mistrust in the relations between science and philosophy at the time. This poor state-of-affairs was blamed on the *Naturphilosophie* proposed by Schelling and Hegel, which was dishonouring, in Helmhotz' opinion, both the positive results achieved by the natural sciences and by Kant's philosophy. Helmholtz advocated accordingly a comeback to Kant's work. Giving center stage to the problem of knowledge was then seen as an effective return to Kant. According to Friedman, the birth of the scientific philosophy ideal, eminently understood as epistemology, can be dated to this moment in time, as recorded by the journal *Vierteljahrsschrift für wissenschaftliche Philosophie* (published in 1877). However, despite placing the origins of the scientific philosophy in the rejection of the *Naturphilosophie*, with an emergent neo-Kantianism as background, Friedman (1992) also admits that the philosophers of nature were trying to give an intellectual framework to the scientific achievements of their time, which involved sensational novelties in the area of chemistry, electricity, magnetism or physiology.

According to Friedman's account, another milestone in the relations between science and philosophy, as well as in the project for a scientific philosophy, is the commemoration of Hermann von Helmholtz' centenary from his birth. In 1921, Moritz Schlick presented at the University of Berlin a memorial address titled 'Helmholtz als Erkenntnistheoretiker,' as a tribute to the great scientist who had largely contributed to the development of psychophysiology, electrodynamics, and geometry, in addition to his studies on acoustics and his own formulation of the conservation of energy principle. One should also mention that Schlick obtained his PhD in Physics under the supervision of Max Planck, yet he soon preferred philosophy as a career, writing works such as *The Essence of Truth in Modern Logic* (published in 1910). His profile is therefore that of a scientific philosopher or a philosophical scientist, solidly trained in physics and interested in epistemological issues. Schlick discussed Einstein's relativity from a philosophical point of view (Engler and Renn, 2018), and earned the Chair for the Philosophy of Inductive Sciences of the University of Vienna, a position formerly occupied by outstanding personalities such as Ernst Mach or Ludwig Boltzmann. He then became one of the leaders of what is usually known as the Vienna Circle of logical positivists.

Schlick notably claims philosophy to be scientific in that science turns out to be philosophical. Within Gary Gutting's classification, Schlick would belong to Kantian criticism: science provides indeed first order knowledge, but it is in philosophy where the necessary conditions which make scientific knowledge possible are revealed. While assuming the validity of science, philosophy is supposed to uncover transcendental truths. Such a season of philosophical optimism, granting a privileged status to philosophy, will not last very long though. After having assimilated Gottlob Frege's and Bertrand Russel's mathematical logic, Ludwig Wittgenstein is seen as delivering the *coup de grâce* to every understanding of philosophy similar to the one advocated by Schlick. In his *Tractatus Logico-Philosophicus*, Wittgenstein (1922, 4.11–4.112, p. 44) writes:

> The totality of true propositions is the total natural sciences (or the totality of natural sciences).
> Philosophy is not one of the natural sciences.
> (The word "philosophy" must mean something which stands above or below, but not beside the natural sciences).
> The object of philosophy is the logical clarification of thoughts.
> Philosophy is not a theory but an activity.
> A philosophical work consists essentially of elucidations.
> The result of philosophy is not a number of 'philosophical propositions', but to make propositions clear.

Little space then remains for the view that considers scientific knowledge as being based on fundamental philosophical principles.

Before closing this brief summary on the scientific philosophy ideal and the interweaving between philosophy and science, it is worth mentioning that Michael Friedman also pays attention to Rudolf Carnap, another member of the Vienna Circle. The ideal of a scientific philosophy was indeed damaged after Wittgenstein's tenets. However, Carnap was trained in those natural sciences accounting for the ultimate frontier of all legitimate epistemology. In particular, he studied physics and philosophy at the University of Jena. He knew very well what natural sciences could or could not bring, and he was also an expert on Russell's, Whitehead's, and Frege's works. Carnap's first great philosophical contribution, *Der logische Aufbau der Welt* (published in 1928), was praised for being the best expression at the time of a philosophy built with the accuracy of mathematical logic. Carnap tried to culminate the project of conceiving and executing an epistemology built on the solid tradition of exact and formal sciences. This endeavour was further developed in his *Logical Syntax of Language* (published in 1934). In this vein, scientific philosophy is understood as a science of logic. From this point of view, to elucidate the ontology of science is equivalent to elucidating the language in which science is expressed, based on the assumption that language represents logical invariances susceptible of being captured in an epistemological analysis. As opposed to Wittgenstein's early mysticism around the ineffable, Carnap advocated a "*perfectly precise and rigorous system of logico-mathematical propositions*" (Friedman, 2001, p. 16). To summarize, Carnap would agree with Kant that philosophy, understood as a kind of logical analysis, is a meta-science which eventually reveals the conditions of possibility of scientific knowledge. However, he would go as far as to claim that philosophy is even a special branch of science, which allows to discern the formal or *a priori* aspects of science, as opposed to the rest of specialities which are essentially involved with the articulation of results produced by empirical research.

Of course, many other references may be added regarding the history of the ideal of a scientific philosophy, such as Frege's *Begriffsschrift*, Russell's and Whitehead's *Principia Mathematica*, Hans Reichenbach's *The Rise of Scientific Philosophy* (published in 1951) or the aforementioned neo-Kantian contributions provided by Hermann Cohen, Paul Natorp, and Ernst Cassirer. In Michael Friedman's (2001) historical reconstruction, the ideal of a scientific philosophy ends with Kuhn, for whom science, the same as philosophy, would not display a privileged epistemological position and both science and philosophy succumb under the movement of their own history.

Moreover, it can be argued that Husserl's phenomenology also belongs to this complex movement of producing a scientific philosophy, presented in terms of a new beginning without prior assumptions. Gary Gutting seems accurate when he places Husserl's phenomenology among the proposals which consider philosophical truths as superior to scientific truths, since Husserl's philosophy aims to provide a more fundamental vision than that of the particular sciences. In this regard, Blumenberg (1986) points out that the ideal of scientificity in phenomenology is indeed an illusion made at the cost of eliminating the contingency of life and history. According to Blumenberg, the supposed origins of the phenomenon—the touchstone of scientificity in phenomenology—is not revealed in the phenomenological reduction of the world, that is to say, in reducing the phenomenon to the evidence of intuition. Rather, the historicity of existence is the very starting point of any philosophical reflection. Blumenberg considers that historicity and contingence call into question the aim of transcendentality. The temporality of the epistemological subject is irreducible and undermines the Husserlian project of turning philosophy into a strict science. In order to develop such a project, a transcendental subjectivity is needed, a unifying and systematizing element of the whole phenomenological building, able to eliminate the concretion of existence and to produce a pure conscience free of history and opened to the essence of phenomena. Blumenberg (1997a) ironically likens the phenomenologist and their pursuit of essences to a first human being who sees the world as on the first day.

3 Hans Blumenberg's History and Philosophy of Science

As already mentioned, Blumenberg's insights on the history and philosophy of science follow in the wake of Husserl's attempt at building a scientific philosophy based on a new theory of knowledge.[2] This circumstance clearly positions Blumenberg's works within the tradition of the continental history and philosophy of science. Nothing seems more distanced, though, from the idea of a philosophy as a strict science than Blumenberg's *Umwege*, i.e., his brilliantly written digressions, anecdotes, and biographies of metaphors (Koerner, 1993).

In spite of his close ties to phenomenology (Blumenberg, 2018; Fragio Gistau et al., 2019), Blumenberg's philosophical starting point are the scandals of contingency and facticity, mixed with conceptual history (*Begriffsgeschichte*), hermeneutics, and anthropological philosophy, among other trends. Blumenberg's theoretical approach may be taken as a phenomenological anthropology of history, or simply, as a phenomenology of history, in line with his mentor Ludwig Landgrebe (1968). Indeed, Blumenberg practices an anthropological turn with the help of the contributions of authors such as Gehlen, Plessner, Alsberg, and Cassirer. As a result, a prominent position is granted to the plural expressive configuration of the 'life-world' (*Lebenswelt*) (Blumenberg, 1972, 2010a; Husserl, 1970), focusing on the evolution and diversification of absolute metaphors and myths across Western history.

The main ideas featured in Blumenberg's work include, among others, the absolutism of reality, the critical revision of phenomenology, and the intellectual history of science and technique. However, for the early Blumenberg, the model of scientific knowledge is not mathematics—like for Husserl—nor the natural and formal sciences—like for the logical positivism or neo-Kantianism—but rather astronomical knowledge (Fragio Gistau, 2023). Astronomy is able to show the facticity of human existence, our vulnerability and marginal position in the universe. While Husserl's mathematics and his *Philosophie als strenge Wissenschaft* (published in 1911) put us away from life, astronomy paves the way for a new human self-understanding. Furthermore, Blumenberg complements astronomy as a model of scientific knowledge with anthropology. The anthropological semantics of the cosmos is the topic of books such as *The Genesis of the Copernican World* (published in 1975), *Lebenszeit und Weltzeit* (published in 1986), and *Die Vollzähligkeit der Sterne* (published in 1997). The absolutism of reality notably appears in the starry heavens, in the open sea, in the shipwreck or in the steppes of anthropogenesis (Blumenberg, 1996, 1997c, 2006, 2010b).

---

[2] Similar to other cases, Husserl presented his scientific philosophy in terms of a new beginning without previous assumptions: "*For Descartes and Bacon the problem of the 'beginning' was simpler, post-scholastic: for both of them, all preceding history yielded nothing but an enormous ballast of prejudices, which could easily be thrown overboard, leaving nothing but the task of a new construction without previous assumptions. Husserl had also made this pathos of the radical beginning his own, connecting it to the methodical demand of 'free variation', for which the complex of representations of a world constituted an available and manageable resource at will*" (Blumenberg, 1999, p. 25, a. trans.).

However, Blumenberg does not seem particularly interested in the technical discussions on the size of the Universe, the determination of the Earth's weight, the parallax, or the cosmic rays. He is rather interested in everything which lends itself to a philosophical anthropology of science, namely what suggests the excess of an inhospitable universe and our marginal position in it. He focuses then on problems such as why Rhetico—the genuine heir to the Copernican astronomy—ended up erecting an obelisk to observe the sky and recovering the old-fashioned Egyptian cosmology; or Kant's pre-critical doubts whether the Earth's surface is ageing (Blumenberg, 1986, 1987). For Blumenberg, science diminishes the meaning of reality, showing how the universe remains impassive to human hopes. At the same time, scientific culture brings the opportunity to produce a more human world.[3] For instance, the telegraph laying along the North Pole, which breaks the ice's silence, symbolizes mankind's wish of escaping from a destruction coming from an indifferent reality (Blumenberg, 2020a; González Cantón, 2004).

From early books like *Paradigmen zu einer Metaphorologie* (published in 1960) to later ones such as *Begriffe in Geschichten* (published in 1998) and *Beschreibung des Menschen* (published in 2006), philosophical anthropology becomes increasingly prominent in Blumenberg's works. He notably applies to history the principle of insufficient reason and the theory of unconceptuability (Blumenberg, 1997b). This clearly affects also Blumenberg's views on science and his position in relation to history and philosophy of science. From an intellectual history of astronomy he moves to a hermeneutics of the life-world, expressed in fables and anecdotes. Blumenberg's late production, in particular, features less monumental works accounting for the history of epochal thresholds and absolute metaphors in terms of exercises in self-understanding. Rather, self-understanding is made explicit in the return to the *Lebenswelt* and its symbolic manifestations.

Throughout his production, though, Blumenberg's view on science remains tied to the critique of phenomenology. He argues that the facticity of existence replaced the *factum* of science (Blumenberg, 1986). In the later phase of Blumenberg's thought, science is mainly confined to the life-world, and its preeminent function is not to display the absolutism of reality, but the obviousness of the *Lebenswelt*. An amendment is therefore introduced to his early understanding of science. Science is no longer the main stage for the revealing of a phantasmagoric indifferent universe. On the contrary, science, together with technology, makes it possible to endow the world with its comfortable obviousness (Blumenberg, 1998).

[3] Villacañas Berlanga has summarized this point as follows: "science produces self-assertion […] by way of previously producing humiliation" (2017, p. 214).

4 Blumenberg Revisits Cassirer

Besides the philosophical anthropology of science in Blumenberg's monumental works, or his conceptualizations along the lines of Husserl's phenomenology, one of the great novelties provided by Blumenberg's contributions to the history and philosophy of science may be found in his response to Cassirer's work and in his consequent avoidance of the "overwhelming determinism of a world of possible scientific objects" (1999, p. 128, a. trans.).

In his famous acceptance speech for the Kuno Fischer Prize, awarded by the University of Heidelberg, Blumenberg (2022a) outlines an original philosophy and historiography of science. He points out that neo-Kantianism, and Cassirer in particular, would see in the history of science the triumph of a progressively developing reason, and the genuine *telos* of human knowledge in which all form of epistemological historicity is overcome (Blumenberg, 2010c). For the neo-Kantians, scientific experience is the reference model of human knowledge. The hegemony of science would have accordingly no epistemological limits, even though Kant himself points to them. A scientific *factum* is then assumed to be the only possible foundation of philosophy and the model for its conditions of possibility (Villacañas Berlanga, 2017).

In this sense, it is worth adding that Cassirer is usually considered the key figure in the early neo-Kantian history of science. One of the main supporters of a history of epistemology faithful to the neo-Kantian Marburg school, Cassirer is known for his Platonic interpretation of Galilean science as mathematisation of nature, as conveyed in his ground-breaking work on the scientific culture of the seventeenth century, *Das Erkenntnisproblem in der Philosophie und Wissenschaft der neueren Zeit* (published 1906–1957). According to Cassirer, behind the historical modifications of scientific thought there are formal structures that may remain unchanged for long periods of time. In particular, the modern understanding of nature features the success of the concept of mathematical function over the substantialist Aristotelian metaphysics. This idea was already anticipated in Hermann Cohen's *Das Prinzip der Infinitesimal-Methode und seine Geschichte* (published in 1883), and further developed in Cassirer's *Substanzbegriff und Funktionsbegriff* (published in 1910). According to the early Cassirer, the history of science is a slow and endless process of improvement of our logical and formal tools as well as of our understanding of nature.

However, Cassirer's later philosophy of symbolic forms is based on a historicization of epistemology and shows that science cannot account for all the expressive forms of consciousness, much less can it do justice to those paths that do not belong to the triumphal development of scientific knowledge. The philosophy of symbolic forms—like Husserl's *Lebenswelt*—aspires to recover the ordinary experience, that is to say, all that which cannot be easily reduced to the determinations of scientific experience and which, on the contrary, allows the latter to be viable:

> It was henceforth to be impossible to make scientific progress the indicator of all differentiation of consciousness, as is proved by the comprehensive reappraisal of findings made by

> linguistics, ethnography and the history of religion. History appears as but one of the relations in which what is foreign may be perceived as something potentially our own. Time and space are accorded equal rank as the dimensions of this fundamental experience, from which arises the task of seeing and assessing what is foreign not by measure of what we consider our own, nor the past by measure of the present (Blumenberg, 2022a, pp. 217–218).

This not only means the careful recovery of the gaze of the other as distinct from oneself, but also to avoid the understanding of the history of science as a teleology towards present-day successes, relegating to obscurity and the condition of mere instruments all those previous formulations that do not reach our own degree of epistemological perfection (Villacañas Berlanga, 2017). Cassirer himself would avoid this excessive ambition of science, which uses the past as a tool subordinated to scientific progress. In contrast to this vision that understands science as the supreme form of reason, the philosophy of symbolic forms gathers the humble expressions of the life-world, with the aspiration of doing justice to them. According to Blumenberg, Cassirer would overcome the teleological understanding of science in his praxis as a historian. Cassirer introduces a historiographical *ethos* able to compensate the excesses of science's teleologism, that is to say, against the mediatization of history in the service of the present. Books such as *The Individual and the Cosmos in Renaissance Philosophy* (published in 1927) and *Die platonische Renaissance in England und die Schule von Cambridge* (published 1932) bear testimony to Cassirer's interest for those cultural processes which do not fit in the shining development of scientific progress, which have been historically interrupted, and are a monument to failed persons and extinct perspectives (Villacañas Berlanga, 2017).

Cassirer declares a historian's ethics inconsistent with his deepest neo-Kantian epistemological convictions, but nevertheless necessary to repair the injustice practiced in the teleological historiography of science. This historian's ethics extends the Kantian categorical imperative to the past, and commits not to treating the dead merely as a means to an end, but always as an end in itself:

> What remains to be learned from Cassirer resides precisely in what he did not succeed in yet is discernible as an urgent impulse throughout his life's work and beyond: not to make the history of philosophy of the sciences, of the systems of symbolic forms subservient to the self-confirmation of present states, to the criterion of success – or indeed to their relevance to consciousness formation. Nowhere did Cassirer let us know whether it was as a Kantian that he resisted – successfully or not – the imperative to use humanity, even in a single person, as a means to an end, even to that of functionalizing history with a view to the present's need for topical relevance (Blumenberg, 2022a, p. 219).

It may be argued that this 'ethic of knowledge' (González Cantón, 2004) is also operating in Blumenberg's work, in contrast to the teleological view of history and philosophy of science.

5 Blumenberg's Critique of Science

Blumenberg's work establishes that philosophy as a strict science or, by extension, science as the only valid and legitimate epistemic form of all possible experience, is inspired by a framework of quite problematic assumptions. Science may be understood as a mimetic project of perfect adjustment between the world and the system of representations, which also has the added benefit of technological production. However, the mimetic movement of science, presumably capable of leaving nothing out of its consideration, is an endless task and requires an infinite time that exceeds the duration of human life (Blumenberg, 1986; see Koerner, 1993), and is therefore unfair to the finite *Lebenswelt* that motivates and enables it (Villacañas Berlanga, 2017). In this sense, Blumenberg points out a dialectic of science which consists in decoupling the personal experience of the world—a singular, historical and unique one because of its finitude—from the scientific experience of the world—a collective, impersonal and transhistorical one—which quickly devours resources and life-time:

> The institutional nature of science lies precisely in the fact that it is not affected by the vital processes of the subjects who are involved in it. It follows that the plurality of its bearers is irrelevant to its theoretical function. The process runs as if there was one single identical subject in its motives and ends. The pure divergence between the life-time and the world-time becomes a functional contraposition, due to the fact that the demands of objectivity force the demands of subjectivity into asceticism. The pretension to reach the dimensions of the world ultimately presupposes, as a last consequence, the renunciation of maintaining the dimension of the life-time as a criterion of the expectations of meaning. The success of the quest for knowledge goes hand in hand with the renunciation of meaning (1986, p. 173, a. trans.).

Science is made by individuals, but it emancipates from them, acquiring the status of an entity that seems to subsist on its own, and demands collective, intergenerational attention, as a kind of transhistorical lonely subject. In short, science, pushed out of its limits, unfolds in a dialectic between progress and the failure of progress:

> The growth of science, of its exertion in work, life achievement, and institutions, as well as in its economic means, is, when closely observed, the fulfillment of the Cartesian ideal according to the structure of context: the achievement of theory encompasses the singularity of lives. However, the notion that science must constantly grow, not only in the yield of its results but also in the effort of its actors, would, from the perspective of its early programmatic thinkers, not be progress but failure (Blumenberg, 2010b, p. 151).

The promises of scientific knowledge inevitably lead to disappointment. Everything seems to be reducible to science. The totality of facts is apparently saved from oblivion by means of a systematic reason providing one single meaning to all human achievements (Adams, 1991; Villacañas Berlanga, 2001). The rhetoric of science wants scientific knowledge to be seen as the ultimate horizon of meaning (Blumenberg, 1999). Accordingly, the scientific construction is considered as forever enduring and bound to house a fully realised humanity.

However, in compensation for this undervaluing of life, science constantly renews its promise of intra-worldly salvation, and with it a form of asceticism, even if its results are understood as provisional and perfectible. Precisely because of the provisionality of science, the definitive achievements of scientific activity, together with those yet to come, are projected into an indeterminate future, but one that is not, at least in principle, unattainable. The voracity of science is thus justified, and opens up an indefinite waiting period, marked by the asceticism of patience (Villacañas Berlanga, 2001).

5.1 *The Ambiguity of Science*

After this brief review of Blumenberg's critique of science, it might be already clear that science is a quite ambiguous topic in his work. This ambiguity is twofold and is expressed by two different orders of ideas. As to the first: while science can unveil the meaninglessness and indifference of reality, it lacks the legitimacy needed to become the final stage of philosophical thought. According to Blumenberg, science assumes the world to be internally coherent. But while it demands such state of affairs as a condition of possibility, it typically rejects any theological or metaphysical foundation to guarantee that reality and scientific representation fit one another (see González Cantón, 2004). As a result the development of science coincides with reality being immanently unveiled and the proof of its utter purposelessness being provided. However, this contrasts with the transformation and substitution of reality thanks to technological production, i.e., through the realization of a more human world inspired by such internal coherence. This first ambiguity is also conveyed by Blumenberg's principle of insufficient reason, based on Leibniz' philosophy. For Leibniz, the world is just one of an infinity of possible worlds, but it is not contingent since there is a sufficient reason that justifies its existence. As far as Blumenberg is concerned, the world is also one of the possible worlds, but it is not necessary, since there is no "world formula" or philosophy of history (2022b, p. 98ff.). In other words, the internal coherence of reality is not sufficient to save reality from contingency, but how then to justify the performances of science that Blumenberg seems to respect? On this matter, Blumenberg's position is quite ambivalent[4]:

[4]Ambiguity is also constitutive of the principle of insufficient reason: "But the principle of insufficient reason is not to be confused with a demand that we forgo reasons, just as 'opinion' does not denote an attitude for which one has no reasons but rather one for which the reasons are diffuse and not regulated by method. One has to be cautious about making accusations of irrationality in situations where endless, indefinitely extensive procedures have to be excluded; in the realm of reasoning about practical activities in life, it can be more rational to accept something on insufficient grounds than to insist on a procedure modeled on that of science, and it is more rational to do this than to disguise decisions that have already been made in arguments that are scientific in form" (Blumenberg, 2020b, p. 198).

> The millennium that is now drawing to a close has been full of talk about the demand for meaning and the loss of meaning, about seduction through offers of meaning and dismay at their failure, but it has never drawn inferences from what was lost to what was demanded, nor has it described the withdrawal even in bare outline. Orientation in the face of such helplessness may therefore be sought in a typology of the forms in which meaning had once been realized, projected, or already forgone. There can be no question of restoring the world as the readability of friendly or grudging, threatening or propitious messages for humankind. But the privileging of a unique kind of experience, one aimed not so much at making us feel at home in the world as at placing it at our disposition, can be recognized as non-self-evident and historically contingent, albeit inescapable. This approach should not be chalked up to the "critique of science", since any deprecation of the unsurpassable contribution to our lives made by modern science is both unwanted and abhorrent to me; to flirt with despising it is itself despicable. To say that it does not exhaust the sum of the possible, however, is more than a mere platitude (2022b, pp. 3–4).

The second order of ideas can be summarised as follows: science is not the place of cognitive freedom, but the very limit of thought. In order to understand this insight, we need to make some preliminary considerations. Prior to Blumenberg's intellectual development, Heidegger publishes *Sein und Zeit* as an *a priori* account of concrete existence. Said *a priori* claims cancel the epistemic historicity of the categories used, turning Heidegger's famous book into a philosophical contribution presumably situated beyond history (Friedman, 2000). Against the backdrop of a missing epistemic historicity, the entity with privileged access to being, i.e., *Dasein*, is to be considered from the ontological primacy of time, insofar as it is a 'being-towards-death.' Consequently, the existential analytic of *Dasein* is an *a priori* hermeneutics of death expressed in a timeless language.

Differently, Blumenberg sees the facticity of existence as also expressed in *logos*. This amounts to saying that also historical symbolic-cultural configurations reveal the analytics of *Dasein* (Jonas, 1958). Metaphorology captures the finitude and temporality of *Dasein* at the semantic level. In this sense, metaphorology also involves a hermeneutics of facticity with a language that is not outside history and encodes the spontaneous exercise of human self-understanding (Blumenberg, 1997a). Moreover, the proper place to explore existential phenomenology is in the production of symbolic forms linked to the *Lebenswelt* and not in an *a priori* analysis of human facticity, since this last kind of approach distort both contingency and historicity. The ontological pre-eminence that Heidegger reserves to *Dasein*—the entity that possesses, so Heidegger, the privilege of access to being as totality—is now considered in the historicity of symbolic configurations such as metaphors, myths, or rhetoric.

Metaphorology captures the creative freedom of consciousness as is shown by the history of metaphors and symbolic forms. Whereas Heidegger sees in the *Dasein* the entity with ontological pre-eminence for accessing the historicity of being, Blumenberg identifies in the evolution of metaphors and myths the privileged epistemological place to establish a new anthropological phenomenology of history. The temporality of human beings is concretised in their freedom to create meaning. The evolution and disappearance of metaphors, that is to say, the 'being-towards-death' of metaphors, yields a unique understanding of historicity: it is both an expression of human finitude and of human freedom. This is why metaphorology is opposed to philosophy as a strict science, inasmuch as the latter implies a deprivation, a reduction of the plurality of horizons of meaning to a single horizon: that of scientific knowledge.[5] In contrast to the rigour, uniqueness and exhaustiveness of science, Blumenberg proposes digressions and detours: "excursuses demand a degree of freedom that cannot be tolerated in the discourse of rational thought" (1991, p. 231).

The aporia of freedom brought to representation consists in the fact that science "does not achieve what our wishes and claims had expected of it, but what it does achieve is essentially unsurpassable" ["*aber was sie leistet, ist nicht wesentlich überbietbar*"] (Blumenberg, 1997b, pp. 73–74).[6] Science then establishes itself as a limit to thought, and enters into confrontation with the endless exercise of human self-understanding accounted for by metaphorology. To put it otherwise: the hermeneutics of facticity may be applied at the level of *Dasein*—which Blumenberg does not seem to be particularly interested in—as well as at the level of the life and death of symbols and metaphors—a key to understand Blumenberg's work. In this train of thought, science is presented as an insurmountable barrier that puts an end to the movement of freedom, that is to say, the "freedom of deviation" (Blumenberg, 1991, p. 231). According to this vision, science blocks the Husserlian free variation applied by Blumenberg to intellectual history,[7] and demands the conceptual necessity and representational systematicity that metaphorology refuses to grant in its hermeneutics of the life-world. To use Feyerabend's terminology,[8] we may refer to a heuristic anarchism of metaphorology that is forbidden by science because of the

[5] Blumenberg has eloquently expressed it in a commentary on Fontenelle and the Enlightenment: "But the successful Enlightener—Fontenelle—nevertheless seems not to be entirely insensitive to the fact that the legend of the oracles' falling silent had offered people's minds something that, after its destruction by science, would not be easy to replace" (1985, p. 107).

[6] A variation of this claim may be found in *Work on Myth*: "Myth is made obsolete by what comes after it; science cannot be made obsolete, however much, in each of its steps forward, it itself makes the preceding steps obsolete" ["Der Mythos wird überlebt durch das, was nach ihm kommt; Wissenschaft ist unüberlebbar, so sehr sie selbst in jedem ihrer Schritte die vorhergehenden überlebt"] (Blumenberg, 1985, p. 50).

[7] One remarkable example of free variation is Nietzsche's reception of the paradise expulsion myth: "Nietzsche says that it was God himself *'who at the end of his days' work lay down as a serpent under the tree of knowledge: thus he recuperated from being God'*. As a snake he not only takes a rest after his work of creation, he makes himself into the principle of evil" (Blumenberg, 1985, pp. 176–177).

[8] The affinities between Blumenberg and Feyerabend are remarkable. The familiarized reader can recognize an anticipation of Feyerabend's positions in quotes from Blumenberg such as: "The history of science has almost always joined in the mockery of those who were proved right. But what about Sizzi, Magini, and Cremonini, of whom we so often hear as not wanting to look through Galileo's telescope or, when they did, claiming to see reflections rather than the moons of Jupiter? […] To the history of philosophy, the character of Simplicio in Galileo's *Dialogue Concerning the Two Chief World Systems* is no longer just an amusing personage but a polemical caricature. The history of science, which by its subject is tied to the notion of success, finds itself in awkward straits when it comes to according respect to those who have fallen into obscurity" (2022a, p. 221).

requirements of an unequivocal epistemological and ontological articulation, which is of course absent in the metamorphoses of myths, metaphors, anecdotes, and fables that make up Blumenberg's work.

These two main insights have a number of implications. For instance, science as a limit to thought may also involve a limit to action (Villacañas Berlanga, 2001); its potentiality empties and fills the world with meaning.[9] Science aspires to close the totality of what can be experienced, but in doing so it only manages to expand that totality; its increasing imposition consolidates the rhetoric of its legitimacy, but at the same time it makes the legitimacy of other forms of experience clear.

6 Book Structure

This volume delves into these outlined topics through contributions written by specialists in Hans Blumenberg's work. It collects the main outcomes and later developments of the international conference *How to Do History and Philosophy of Science with Hans Blumenberg?*, supported by the German Science Research Foundation (DFG). The conference was co-organized by prof. Dr. Cornelius Borck (IMGWF—Universität zu Lübeck), prof. Dr. Alberto Fragio (Universidad Autónoma Metropolitana, Mexico City), Dr. Martina Philippi (Universität Paderborn) and prof. Dr. Josefa Ros Velasco (Universidad Complutense de Madrid). It was aimed both to commemorate the 100th anniversary of Hans Blumenberg's birth and to pay attention to his contributions on history and philosophy of science. On the whole, the book takes an in-depth look at Blumenberg's contributions to continental history and philosophy of science.

This first section, titled "A Philosophical History of Sciences," introduces the reader to Blumenberg's approach to science. Blumenberg's history and philosophy of science evolved towards a historical phenomenology of scientific knowledge through the study of the historical and cultural backgrounds that made the emergence of Modern science possible. Chapter 2, "Hans Blumenberg's History of Astronomy: An Overview," by Alberto Fragio (Universidad Autónoma Metropolitana, Unidad Cuajimalpa), offers a general picture of Blumenberg's interest in the history of astronomy. In the 1950s and 1960s, Blumenberg published a series of articles on modern astronomy subsequently culminating in four main books: *Die kopernikanische Wende* (1965), *Die Genesis der kopernikanischen Welt* (1975), *Das Lachen der Thrakerin* (1987) and the posthumous *Die Vollzähligkeit der Sterne* (1997c). The author focuses on this substantial corpus of texts organised around modern astronomy and its continuation in twentieth-century astronomy. Chapter 3, "On Blumenberg's Mathematical Caves, or: How Did Blumenberg Read Wittgenstein's Remarks on the Philosophy of Mathematics?," by Michael Friedman

[9] Blumenberg finds one of the most extreme examples in Schrödinger's *What is Life?* (published in 1943). The manipulation of the genetic code could make rewriting human nature possible. For further details on the topic, see Blumenberg, 2022b).

(Cohn Institute, Tel Aviv University), analyzes Blumenberg's position with regard to history of mathematics through a close reconstruction of the reception of Wittgenstein's works.

Section 2, "Phenomenology as Anthropology," pays attention to Blumenberg's interest in the connection between phenomenological and anthropological topics, further culminated in his phenomenological anthropology famously outlined in the posthumous book *Beschreibung des Menschen* (2006). Chapter 4, "Crisis and Restitution. Phenomenological Subjectivity as Root of Blumenberg's Description of Historical Change," by Pedro García-Durán (Universitat Jaume I of Castello, Spain), focuses on Blumenberg's reception of transcendental phenomenology applied to the analysis of historical changes. The phenomenological concepts and notions connected to subjectivity, which underpin Blumenberg's understanding of history and historicity, are investigated within this context. The main goal is here to show how said notions establish a key linkage among the 'concept of reality' (*Wirklichkeitsbegriff*), the 'lifeworld' (*Lebenswelt*), and the phenomenological anthropology. Chapter 5, "Lifeworld and Historicity: The Fate of Phenomenology," by José Luis Villacañas (Universidad Complutense de Madrid), delves into Blumenberg's phenomenology of the lifeworld in relation to Anthony J. Steinbock's book *Home and Beyond: Generative Phenomenology after Husserl*. It examines how the internal evolution of phenomenology leads to anthropology as shown by Blumenberg's *Theorie der Lebenswelt* (2010a). This allows to shed light at the same time on Steinbock, Husserl and Blumenberg. Villacañas concludes by pointing out the programme of a situational phenomenological anthropology. Chapter 6, "Defense of 'Expanded Reason' in Hans Blumenberg: The Case for the Science of Boredom," by Josefa Ros Velasco (Universidad Complutense de Madrid), places the philosophical anthropology of boredom as a tool for exploring the limits of scientific rationality and the need for an expanded human reason.

Section 3, "Hans Blumenberg's History and Philosophy of Technology," is devoted to Blumenberg's intellectual history of technology, as evinced by texts written in the 1950s and 1960s. In these early contributions, Blumenberg envisioned a connection among the philosophy of technology, phenomenology and intellectual history, further developed in his posthumous books *Geistesgeschichte der Technik* (2009) and *Schriften zur Technik* (2015). Chapter 7, "Technique, Technology, and Theory. Blumenbergian Perspectives in Philosophy of Culture," by Christine Blättler (Christian-Albrechts-Universität zu Kiel), looks at Blumenberg's understanding of theory from the viewpoint of cultural techniques involving observation, description and meaning production. This approach overcomes the classical view of theory as contemplation and emphasises theory in its intervening dimension. It shows how theory is not only intertwined with technology but also with grand narratives, apocalyptic imaginaries and encompassing cultural backgrounds. Chapter 8, "Nobody's Driving. Reality and Technology According to Hans Blumenberg," by Gonzalo Forsberg (Leiden University), is devoted to Blumenberg's understanding of technology in its transformative effect upon reality. With this aim, the author undertakes an extensive review of Blumenberg's texts on technology and its connection with phenomenology, the concept of reality and the 'life-world' (*Lebenswelt*).

Chapter 9, "Unburdening the World with 'Worlds on Demand.' A Time Capsule from the Legacy" by Martina Philippi (Universität Paderborn), connects the scientific-historical entanglements described in the previous chapters to the new field of 'world-simulating' technologies. The contribution builds a bridge from central ideas in Blumenberg's work to current phenomena of consumption and digitalization and their consequences.

References

Adams, D. (1991). *The divine art of forgetting: Aesthetic distance in Benjamin, Blumenberg, and Pynchon*. City University of New York.

Babich, B. (2019). Early continental philosophy of science. *Articles and Chapters in Academic Book Collections, 31*, 263–286.

Blumenberg, H. (1965). *Die kopernikanische Wende*. Suhrkamp.

Blumenberg, H. (1975). *Die Genesis der kopernikanischen Welt*. Suhrkamp.

Blumenberg, H. (1972). The life-world and the concept of reality. In L. E. Embree (Ed.), *Life-world and consciousness: Essays for Aron Gurwitsch* (pp. 425–444). Northwestern University.

Blumenberg, H. (1985). *Work on myth*. MIT Press.

Blumenberg, H. (1986). *Lebenszeit und Weltzeit*. Suhrkamp.

Blumenberg, H. (1987). *The genesis of the Copernican world*. MIT Press.

Blumenberg, H. (1991). Pensiveness. In D. Adams (Ed.), *The divine art of forgetting: Aesthetic distance in Benjamin, Blumenberg, and Pynchon* (pp. 229–237). City University of New York.

Blumenberg, H. (1996). *Höhlenausgänge*. Suhrkamp.

Blumenberg, H. (1997a). *Ein mögliches Selbstverständnis. Aus dem Nachlaß*. Reclam.

Blumenberg, H. (1997b). *Shipwreck with spectator. Paradigm of a Metaphor for existence*. MIT Press.

Blumenberg, H. (1997c). *Die Vollzähligkeit der Sterne*. Suhrkamp.

Blumenberg, H. (1998). *Begriffe in Geschichten*. Suhrkamp.

Blumenberg, H. (1999). *Wirklichkeiten in denen wir leben. Aufsätze und eine Rede*. Reclam.

Blumenberg, H. (2006). *Beschreibung des Menschen*. Suhrkamp.

Blumenberg, H. (2009). *Geistesgeschichte der Technik*. Suhrkamp.

Blumenberg, H. (2010a). *Theorie der Lebenswelt*. Suhrkamp.

Blumenberg, H. (2010b). *Care crosses the river*. Stanford University Press.

Blumenberg, H. (2010c). *Paradigms for a Metaphorology*. Cornell University Press.

Blumenberg, H. (2015). *Schriften zur Technik*. Suhrkamp.

Blumenberg, H. (2018). *Phänomenologische Schriften (1981–1988)*. Suhrkamp.

Blumenberg, H. (2020a). Advancing into eternal silence: A century after the sailing of the Fram. In H. Bajohr, F. Fuchs, and J. P. Kroll (Eds. and Trans.), *History, metaphors, fables: A Hans Blumenberg reader* (pp. 571–580). Cornell University Press.

Blumenberg, H. (2020b). An anthropological approach to the contemporary significance of rhetoric. In H. Bajohr, F. Fuchs, and J. P. Kroll (Eds. and Trans.), *history, metaphors, fables: A Hans Blumenberg reader* (pp. 177–208). Cornell University Press.

Blumenberg, H. (2022a). In memory of Ernst Cassirer: Speech delivered in acceptance of the Kuno Fischer prize of the University of Heidelberg, 1974. *New German Critique, 145*(49), 215–224. https://doi.org/10.1215/0094033X-9439713

Blumenberg, H. (2022b). *The readability of the world*. Cornell University Press.

Domski, M., & Dickson, M. (Eds.). (2010). *Discourse on a new method: Reinvigorating the marriage of history and philosophy of science*. Open Court.

Engler, F. O., & Renn, J. (2018). *Gespaltene Vernunft. Vom Ende eines Dialogs zwischen Wissenschaft und Philosophie*. Matthes & Seitz.
Fragio Gistau, A. (2020). Similarities, differences, and missed connections between Thomas S. Kuhn, Gaston Bachelard and the continental historiography of science. *HoST – Journal of History of Science and Technology, 12*(2), 94–111. https://doi.org/10.2478/host-2020-0016
Fragio Gistau, A. (2023). *Paradigmen zu einer Metaphorologie des Kosmos. Hans Blumenberg und die zeitgenössischen Metaphern des Universums*. Verlag Karl Alber.
Fragio Gistau, A., Philippi, M., & Ros Velasco, J. (Eds.). (2019). *Metaphorologie, Anthropologie, Phänomenologie. Neue Forschungen zum Nachlass Hans Blumenbergs*. Verlag Karl Alber.
Friedman, M. (1992). Transition from the Metapphysical foundations of natural science to physics. In *Kant and the exact sciences* (pp. 213–342). Harvard University Press.
Friedman, M. (2000). *A parting of the ways. Carnap, Cassirer, and Heidegger*. Open Court.
Friedman, M. (2001). *Dynamics of reason*. CSLI Publications.
Friedman, M. (2008). History and philosophy of science in a new key. *Isis, 99*(1), 125–134.
Friedman, M. (2012). Scientific philosophy from Helmholtz to Carnap and quine. In R. Creath (Ed.), *Rudolf Carnap and the legacy of logical empiricism* (pp. 1–12). Springer.
González Cantón, C. (2004). *La metaforología de Blumenberg como destino de la analítica existencial*. PhD Thesis, Complutense University of Madrid.
Gutting, G. (2005). What is continental philosophy of science. In *Continental philosophy of science* (pp. 1–16). Blackwell Publishing.
Husserl, E. (1970). *The crisis of European sciences and transcendental phenomenology*. Northwestern University Press.
Jonas, H. (1958). *The gnostic religion: The message of the alien god and the beginnings of Christianity*. Beacon Press.
Koerner, J. L. (1993). Ideas about the thing, not the thing itself: Hans Blumenberg's style. *History of the Humans Sciences, 6*(4), 1–10.
Landgrebe, L. (1968). *Phänomenologie und Geschichte*. Gütersloher Verlagshaus G. Mohn.
Morris, S. (2020). Scientific philosophy and the critique of metaphysics from Russell to Carnap to quine. *Erkenntnis, 85*, 773–799. https://doi.org/10.1007/s10670-018-0049-x
Villacañas Berlanga, J. L. (2001). Esferas de acción y sistema filosófico. El carácter imprescindible de la metáfora. *Daimon. Revista Internacional de Filosofía, 24*, 111–126.
Villacañas Berlanga, J. L. (2017). Legitimacy, self-affirmation and *deificatio*: The modern age as definitive epoch. In A. Fragio Gistau & J. Ros Velasco (Eds.), *Contemporary approaches in philosophical and humanistic thought* (pp. 197–268). Aracne.
Wittgenstein, L. (1922). *Tractatus Logico-Philosophicus*. Kegan Paul, Trench, Trubner & Co..

Alberto Fragio is full professor in the Department of Humanities at the Universidad Autónoma Metropolitana de México, Unidad Cuajimalpa. He earned his Ph.D. in Philosophy (2007) from the Universidad Autónoma of Madrid (Spain), and in Cultural Science (2011) from Scuola Internazionale di Alti Studi di Modena (Italy). From 2007 to 2010 he undertook a dedicated research project on Hans Blumenberg's history and philosophy of science with the support of the Fondazione Collegio San Carlo (Italy). He has published *Paradigmen zu einer Metaphorologie des Kosmos. Hans Blumenberg und die zeitgenössischen Metaphern des Universums* (Nomos/Verlag Karl Alber, 2023). On this same author, he has co-edited *Hans Blumenberg. Nuovi paradigmi d'analisi* (Roma, 2010), and *Metaphorologie, Anthropologie, Phänomenologie. Neue Forschungen zum Nachlass Hans Blumenbergs* (Freiburg/München, 2019).

Josefa Ros Velasco is a Marie Sklodowska-Curie Actions postdoctoral researcher at the Complutense University of Madrid, where she is the PI of the project "Pre-bored. Well-being and prevention of boredom in Spanish nursing homes" (2021–2024). Previously, she was a postdoctoral researcher for the Postdoctoral Research Fellowships at Harvard University for Distinguished

Junior Scholars program (2017–2021). She is a specialist in Boredom Studies from a multidisciplinary perspective, and the founder and president of the International Society of Boredom Studies. She is the author of the book *The Disease of Boredom* (Princeton University Press, 2025), and editor of books such as *Suicide in Modern Literature* (Springer, 2021) or *Boredom Is in Your Mind* (Springer, 2019). For her academic achievements, she received distinctions such as the Spanish National Research Prize 2022 or the Harvard University Lincoln Book Prize 2019, among others.
josros@ucm.es | @JosefaRosUCM | www.josefarosvelasco.com.

Hans Blumenberg's History of Astronomy: An Overview

Alberto Fragio

Abstract Hans Blumenberg's interest in the history of astronomy can be dated back to his first scholarly works. In the 1950s and 1960s, Blumenberg published several papers on Copernicus on *Studium Generale* and held a conference on the same topic at the *Akademie der Wissenschaften und der Literatur zu Mainz* on April 24, 1964. In addition to these early contributions, one should also mention Blumenberg's introduction to a selection of Galileo Galilei's writings. Thereafter, Blumenberg's main contribution to the history of astronomy was fully developed in four main books: *Die kopernikanische Wende*, *Die Genesis der kopernikanischen Welt*, *Das Lachen der Thrakerin*, and the posthumous *Die Vollzähligkeit der Sterne*. One may also add to the list *Lebenszeit und Weltzeit*, where the mystery of time is addressed by Blumenberg as an attempt to overcome both Heidegger's *Sein und Zeit* and Husserl's *Phenomenology of Internal Time-Consciousness*. In this chapter, I will focus on this substantial corpus of texts on modern astronomy and how Blumenberg's interest extended to twentieth-century astronomy.

In this chapter, I summarize the core of Hans Blumenberg's contributions to the history of astronomy, discussed at length in my book *Paradigmen zu einer Metaphorologie des Kosmos. Hans Blumenberg und die zeitgenössischen Metaphern des Universums* (Fragio Gistau, 2023). It has been written within the framework of the research projects PID2021-128606NB100 and PID2022-141667NB-I00 funded by the Spanish Ministry of Science and Innovation, Government of Spain. I would like to thank Tessa Marzotto Caotorta for the English revision. Any remaining mistakes or flaws are, of course, my sole responsibility.

A. Fragio (✉)
Universidad Autónoma Metropolitana, Unidad Cuajimalpa, Mexico City, Mexico
e-mail: afragio@cua.uam.mx

A. Fragio et al. (eds.), *Hans Blumenberg's History and Philosophy of Science*, Studies in History and Philosophy of Science 63,
https://doi.org/10.1007/978-3-031-78251-0_2

1 The Cosmic Understanding of Reality

Concerning the history of astronomy, Blumenberg investigated what made the Copernican heliocentrism possible and how it triumphed in the Modern Age. This leitmotif is closely related to Blumenberg's theory of Modernity as developed in *Die Legitimität der Neuzeit* (published in 1966) (1985). As is well known, in this famous book Blumenberg suggests that human self-assertion is at the basis of Modern Age, as an overcoming of the Gnosticism implicit in late medieval Nominalism. Blumenberg argues that, given the theological absolutism of the late Middle Ages, human self-assertion was to become a necessary implication of every philosophical system.

The modern self-assertion is interpreted by Blumenberg as a response to humans losing centrality in the cosmos. In his PhD Thesis, Blumenberg already puts this argument forward in polemical response to Heidegger's history of being (Dahlke & Laarmann, 2020). One of the novelties introduced by Blumenberg is the characterization of ancient Greek metaphysics as a cosmological ontology. According to this approach, ancient Greek people lived immersed in a secure, ordered, and familiar reality. The distinctive feature of ancient people's cosmic experience of reality would then be a full trust in being, which gives rise to the idea of a natural congruence between the existing and its representation (Blumenberg, 1947). The ancient confidence in being would soon called into question by the numerous experiences that pointed instead to the strangeness of the real, that is to say, the inhumanity hidden in being. In his *Habilitationsschrift*, Blumenberg (1950) describes this process as the collapse of the universal structure of trust in the world, in close connection to the epistemological issue of certainty (Zambon, 2022). In this regard, Descartes brings absolute certainty into the controllable territory of the cogito as a self-assertion of reason. However, Descartes' contribution also involves a devaluation of the whole Middle Ages philosophy as a mere interim in the history of thought as well as the de-powering of the original Christian-medieval understanding of reality. The Cartesian project of certainty would thus have taken away the meaning and intelligibility of scholastic philosophy, ignoring the specific attempts of the Middle Ages to carry out a project of certainty, albeit one of a theological nature. For the early Blumenberg, the medieval project suffered an internal collapse due to the enormous difficulties faced by the Christian theologians in reconciling the original Greek cosmological ontology with the Christian understanding of reality. As an example of this, Blumenberg points to the increasing rationalization of theology as a way for the renewal of the cosmos through the divine creation of the world. However, Nominalism exacerbated the distancing between humans and God, consummating the loss of the cosmic understanding of reality, already underway under the pressure of the various forms of ancient Gnosticism. The late medieval nominalist God did not guarantee the intelligibility of the world or its epistemological access. Hence Blumenberg's (1950) insistence on the affinities between the Cartesian thought experiment of the *genius malignus* and the nominalist concept of God.

Blumenberg resumes all these arguments in *The Legitimacy of the Modern Age*. In this book, the anthropocentric Stoic teleology stands for the cosmic understanding of reality. Stoicism paradigmatically nurtures the illusion of a human-centered cosmology. According to Blumenberg (1985), in Stoicism the confidence in the world is accompanied by a forgetfulness of transcendence, that is to say, a lack of concern for one's own salvation as typical of life in the ancient world. However, the burden of uncertainty for the Stoics rests on their image of the divinity, that of an intrusive God whose providence causes fear. The God of the Stoics would thus have anticipated both the gnostic God and the late nominalist radicalized version of God. This theological uneasiness is compatible, however, with the reassuring advantages of a geocentric cosmology where the human being is placed in the center of the world as a privileged spectator.

Blumenberg (1985) reconstructs in detail the progressive dissolution of confidence in the cosmos in reference to the problem of the plurality of worlds in Democritus and Epicurus, showing how the idea of plurality ends up creating an untenable and ineliminable insecurity, to the extent that the progressive mistrust of reality leads to the gnostic impugnation of the very concept of cosmos. That is why Blumenberg undertakes a comparison of the epochal crisis in Antiquity and in the Middle Ages. The theological crisis of the late Middle Ages leads to self-assertion as a response to late medieval Nominalism which gives rise to a new historical epoch, i.e., the Modern Age. Within this framework, Blumenberg argues, the ancient cosmos could not be restored in the process of overcoming the gnostic legacy. Blumenberg concludes then that Modern Age is the second and definitive overcoming of Gnosticism, as represented in the overcoming of the unbearable theological absolutism of late medieval Nominalism.

2 Copernican Reform

In works such as *Die kopernikanische Wende* (1965b) and *Die Genesis der kopernikanischen Welt* (1975, 1987b), Blumenberg carefully explores and further develops the anthropological-theological-metaphysical implications associated with the human loss of centrality in the cosmos, in particular, in connection with what he calls the 'Copernican reform.' In short, the heliocentric astronomy advocated by Copernicus is seen as the result of tensions between the medieval Christian tradition—including the creation doctrine, Nominalism, Protestantism, and Aristotelian derivations—and the Stoic and Platonic traditions. Blumenberg notably describes the Copernican reform as a triumph of Renaissance humanism. In this picture, many medieval assumptions are seen as necessary in order to achieve such a reformulation of the cosmos, but at the same time Copernicanism is seen as responsible for the weakening of the medieval understandings of the cosmos. In other words, Copernicus' cosmological reform contributes to the process of destruction of the medieval understanding of the world and, as such, the heliocentric doctrine is perceived as a threat to the continuity of the medieval perspective.

Blumenberg describes the rise of heliocentrism as the production of a new truth on the universe. The modern astronomical reformer is seen as committed to a new truth about the cosmos that affects the medieval order and the preservation of the anthropological prerogatives. Blumenberg deals with this topic by establishing a connection between Copernicus and Galileo, including many other relevant figures. In both cases, Blumenberg's historical reconstruction points out a new sensational astronomical truth. However, while Galileo was initially convinced that the new heliocentric doctrine would have such a compelling persuasive power as to be able to assert its certainty, Copernicus was much more cautious and suspicious about the worldly effects of the new astronomical truth. Copernicus is said to have been well aware of supporting a controversial cosmological truth, although he never doubted that, despite humans losing their central position in the cosmos, heliocentrism was absolutely true. Blumenberg refers to this issue as to the claim of truth of the Copernican reform.

In contrast to the initial reception of heliocentrism and the attempts at epistemological neutralization, Copernicus resisted the instrumentalist downplaying of his findings as a mere astronomical hypothesis of practical utility. According to Blumenberg, the instrumentalist assessment of the Copernican doctrine equals to an attempt to eradicate its claim of truth. Nevertheless, Copernicus is said to have been adamant about obtaining a new truth about the universe. In this regard, Blumenberg pays close attention to Copernicus' exchanges with authors such as Osiander and Melanchthon, supporters of the nominalist paradigm of astronomy.

In this context, Blumenberg presents Galileo's experiences with heliocentrism as a case of de-potentiation of astronomical truth. Based on observations made with the telescope, Galileo believes that the powerful truth emanating from it would be enough to persuade his contemporaries. However, Galileo's experiences prove quite the opposite. Galileo had indeed to face the impotence of cosmological truth and the strong resistance against new astronomical discoveries.[1]

3 The Fall of Thales of Miletus

In close connection with his previous works on the intellectual history of astronomy, Blumenberg also devotes many efforts to the study of the anecdote of Thales of Miletus falling into a well while observing the stars, which brought upon him the laughter of the Thracian maid who accompanied him. Blumenberg wrote two essays on the history of the reception of the Thales' anecdote, "Der Sturz des Protophilosophen" (1976) and *Das Lachen der Thrakerin* (1987a). One may take these essays as a highly-refined continuation of his works on the history of astronomy, but at the same time as a reworking on the existential topics related to the

[1] In this same sense, Blumenberg frequently plays with the contraposition between the Cusan—pre-Copernican—and the Nolan—post-Copernican. For further details, see Blumenberg (1985, esp. part IV).

Heideggerian tradition. The anecdote not only involves the dichotomy between the 'being-present-at-hand' and the 'unavailable being' from an astronomical point of view, but also the 'being-present-at-feet' and its variations. These are *existentialia*—undetected as such by Heidegger—that Blumenberg presents with great eloquence. Trapped between the apparent unavailable and the hidden available, and almost blind to the most immediate, the Milesian ventured into the darkness of the night believing that objects of high rank were achievable and with the aim of examining each of the heavenly bodies in their universality, without ever descending to anything concrete or to discuss what is present at foot. Thus, Thales fell into a well right in front of him. Hence, in the Platonic tradition of the anecdote, the Thracian maid's recrimination: the Milesian wanted to know the things of heaven, ignoring what he had under his nose and feet (Blumenberg, 1976).

The state of affairs 'at foot' is completely absent from the concerns of the astronomer, who turns his attention towards the remote and unavailable. This astronomical interest is suddenly interrupted by the tangible realities extended under his feet that imperiously demanded his attention as the legitimate content of the *Sorge*. Thales falls because while scrutinizing the phenomena of the sky, it is the hidden that remains close to his feet. The Thracian maid then feels authorized to warn her master, referring to the tangible and binding reality of the practicalities of everyday life, while discussing his inability to reach the stars and his unacceptable disinterest towards the terrestrial world.

Thales remains foreign to the objects of the world, which are as strange and distant to him as the celestial bodies to the Thracian maid. The proto-philosopher's curiosity about the stars is clearly incompatible with low, terrestrial matters. The laughter involves consequences not only for what we consider to be remote to humans, but also for all that should be close to him. Thales' concerns led him to lose the soil under his feet.

4 Closing Remarks: Twentieth-Century Astronomy in Blumenberg's *Nachlass*

Along with his works on the history of modern astronomy, Blumenberg had the opportunity to meet with leading physicists and mathematicians of his time such as Carl Friedrich von Weizsäcker and Pascual Jordan (Fragio Gistau, 2023), and he witnessed the remarkable achievements of the development of twentieth-century astronomy. I would like to conclude this brief overview making reference to the extensive collection of newspaper articles on astronomy and cosmology that Blumenberg excerpted from German, Swiss, and French periodicals over more than three decades (1963–1996). This heterogeneous set of materials presents Blumenberg as a good connoisseur of contemporary astronomy, and it can actually serve to clearly illustrate the cosmological turn that this new astronomical knowledge produced in Blumenberg's later work, as shown in the 'astronoetical glosses' edited in his posthumous book *Die Vollzähligkeit der Sternen* (1997; see also Fischer & Gehring, 2011; Zill, 2011). Most of the newspaper articles on astronomy preserved in Blumenberg's *Nachlass* are excerpted from the German newspapers *Frankfurter Allgemeine Zeitung* and *ZEIT*. Only a few come from newspapers and magazines such as the Swiss *Neue Züricher Zeitung*, the French *L'Express* or the German *Der Spiegel*. These articles display the scientific results derived from successive improvements in observational technologies and from the launching of artificial satellites and space telescopes, which ultimately made previously unexplored regions of the electromagnetic spectrum available for investigation. In this respect, the articles preserved in Blumenberg's *Nachlass* bear witness to the move from optical astronomy—based on the visible light spectrum that reaches the surface of the Earth—to the so-called new astronomies (Longair, 2013), which are nowadays used in high-energy astrophysics. The opening up of the electromagnetic spectrum revealed spectacular and unexpected astronomical phenomena such as radio galaxies, quasars, pulsars, neutron stars, and black holes. The ability to make observations from above the atmosphere opened entirely new domains for astrophysical and cosmological research and provided an even more complex and wider view of the Universe (Longair, 2013). The newspaper articles collected by Blumenberg describe the major astronomical discoveries of the '60s, '70s and '80s, variously associated with the emergence of radio astronomy, gamma ray and X-ray astronomies, infrared and ultraviolet astronomies.

In the journalistic reception of these astronomical findings one can identify an 'epochal threshold.' The sudden 'transparency' of the sky becomes an object of experience previously completely inaccessible. The journalistic chronicles ultimately convey a historical shift in the astronomical experience. The Universe—the supreme and critical object of the theoretical faculty of humans and their world—is finally made 'present-at-hand' and available in all realms of experience (Mazzi, 2010; Zambon, 2017).

From this point of view, the 'astronoetical glosses' collected in *Die Vollzähligkeit der Sterne* (1997) can be seen as the philosophical outworking of Blumenberg's interest in twentieth-century astronomy. Blumenberg's 'glosses' are an expression of the contemporary astronomical findings, and they offered new possibilities for human self-understanding. Consequently, what Blumenberg calls 'astronoetics' can be understood as an exercise in the enquiry into human existence from a cosmological perspective. In this regard, Blumenberg is able to witness first hand the new chapters of the work on myth being produced by twentieth-century astronomy. As a *contemplator caeli,* Blumenberg is also able to witness the astronomical reoccupation of myths and metaphors by contemporary astronomy. The astronomical and astrophysical discoveries that occurred during the second half of the century not only allow the survival of the myth, but also lead to its renewal. Through their sensational findings astronomy and astronautics re-mythologized the cosmos. This re-mythologization of the cosmos is closely linked to the exploration of the solar system with probes and satellites. A substantial number of newspaper articles preserved in Blumenberg's *Nachlass*, such as his astronoetical glosses, are related to the achievements of American and Soviet probes. These new 'sidereal messengers' revealed sensational reports about Jupiter and its high-temperature seas of liquid hydrogen and the volcanic landscapes of Io with hills as high as those on Earth; detailed images of the surface and atmosphere of Mars, of the ridges and valleys of Venus, and the pale blue north pole of Neptune; the methane snow of Triton, the rings of Uranus, and Titan as a kind of primitive Earth. However, at the same time, together with the deployment of this celestial spectacle, Blumenberg detects the by-product of an insidious regeneration of the world's enigmas and the revival of the cosmological speculation. While providing the ultimate projectivity horizon, the Universe is seen as unfolding as an original background of inconceivability, that is to say, as '*Unbegrifflichkeit*' (Blumenberg, 2007).

References

Blumenberg, H. (1947). *Beiträge zum Problem der Ursprünglichkeit der mittelalterlich-scholastischen Ontologie*. Christian-Albrechts-Universität zu Kiel.

Blumenberg, H. (1950). *Die ontologische Distanz. Eine Untersuchung über die Krisis der Phänomenologie Husserls*. Christian-Albrechts-Universität zu Kiel.

Blumenberg, H. (1964). Kopernikus im Selbstverständnis der Neuzeit. In *Akademie der Wissenschaften und der Literatur in Mainz. Abhandlungen der geistes und sozialwissenschaftlichen Klasse* (pp 339–368).

Blumenberg, H. (1965a). Das Fernrohr und die Ohnmacht der Wahrheit. In G. Galilei (Ed.), *Sidereus Nuncius (Nachricht von neuen Sternen). Dialog über die Weltsysteme (Auswahl). Vermessung der Höhle Dantes. Marginalien zu Tasso* (pp. 7–75). Insel Verlag.

Blumenberg, H. (1965b). *Die kopernikanische Wende*. Suhrkamp.

Blumenberg, H. (1975). *Die Genesis der kopernikanischen Welt*. Suhrkamp.

Blumenberg, H. (1976). Der Sturz des Protophilosophen. Zur Komik der reinen Theorie – anhand einer Rezeptionsgeschichte der Thales-Anekdote. In W. Preisendanz & R. Warming (Eds.), *Das Komische* (pp. 11–64). Fink.

Blumenberg, H. (1985). *The legitimacy of the modern age*. MIT Press.

Blumenberg, H. (1987a). *Das Lachen der Thrakerin. Eine Urgeschichte der Theorie*. Suhrkamp.

Blumenberg, H. (1987b). *The genesis of the Copernican world*. MIT Press.

Blumenberg, H. (1997). *Die Vollzähligkeit der Sterne*. Suhrkamp.

Blumenberg, H. (2007). *Theorie der Unbegrifflichkeit*. Suhrkamp.

Dahlke, B., & Laarmann, M. (2020). Nachwort der Herausgeber. In H. Blumenberg (Ed.), *Beiträge zum Problem der Ursprünglichkeit der mittelalterlich-scholastischen Ontologie* (pp. 217–230). Suhrkamp.

Fischer, M., & Gehring, P. (2011). Die Kränkung der Venus. Astronomie zwischen Wissenschaftsgeschichte und Astronoetik. *Journal Phänomenologie, 35*, 8–12.

Fragio Gistau, A. (2023). *Paradigmen zu einer Metaphorologie des Kosmos. Hans Blumenberg und die zeitgenössischen Metaphern des Universums*. Verlag Karl Alber.

Longair, M. (2013). The opening up of the electromagnetic Spectrum and the new astronomies. In *The cosmic century. A history of astrophysics and cosmology* (pp. 125–170). Cambridge University Press.

Mazzi, E. (2010). I pensieri astronoetici come laboratorio per un'antropologia sperimentale: la riflessione di Hans Blumenberg sull'impresa spaziale. In A. Fragio Gistau & D. Giordano (Eds.), *Hans Blumenberg: Nuovi paradigmi d'analisi* (pp. 263–299). Aracne.

Zambon, N. (2017). *Das Nachleuchten der Sterne. Konstellationen der Moderne bei Hans Blumenberg*. Wilhelm Fink.

Zambon, N. (2022). Nachwort des Herausgebers. In H. Blumenberg (Ed.), *Die ontologische Distanz. Eine Untersuchung zur Krisis der philosophischen Grundlagen der Neuzeit* (pp. 349–376). Suhrkamp.

Zill, R. (2011). Zu den Sternen und zurück. Die Entstehung des Weltraums als Erfahrungsraum und die Inversion des menschlichen Erwartungshorizonts. In M. Moxter (Ed.), *Erinnerung an das Humane. Beiträge zur phänomenologischen Anthropologie Hans Blumenberg* (pp. 300–326). Mohr Siebeck.

Alberto Fragio is full professor in the Department of Humanities at the Universidad Autónoma Metropolitana de México, Unidad Cuajimalpa. He earned his Ph.D. in Philosophy (2007) from the Universidad Autónoma of Madrid (Spain), and in Cultural Science (2011) from Scuola Internazionale di Alti Studi di Modena (Italy). From 2007 to 2010 he undertook a dedicated research project on Hans Blumenberg's history and philosophy of science with the support of the Fondazione Collegio San Carlo (Italy). He has published *Paradigmen zu einer Metaphorologie des Kosmos. Hans Blumenberg und die zeitgenössischen Metaphern des Universums* (Nomos/Verlag Karl Alber, 2023). On this same author, he has co-edited *Hans Blumenberg. Nuovi paradigmi d'analisi (Roma, 2010), and Metaphorologie, Anthropologie, Phänomenologie. Neue Forschungen zum Nachlass Hans Blumenbergs* (Freiburg/München, 2019).

On Blumenberg's Mathematical Caves, or: How Did Blumenberg Read Wittgenstein's Remarks on the Philosophy of Mathematics?

Michael Friedman

Abstract In the chapter "Im Fliegenglas" from his book *Höhlenausgänge*, Blumenberg refers to Wittgenstein's views on mathematics, logic, and mathematical proof on several occasions. One of Blumenberg's key points concerning Wittgenstein's famous statement that philosophy's aim is to show the fly the way out of the fly-bottle is discussed with a rather surprising example: a geometrical puzzle that involves decomposing a rectangle. Although Blumenberg is not known for developing a comprehensive theory of the philosophy of mathematics or mathematical practices, this chapter analyzes Blumenberg's reflections on mathematics and mathematical concepts, which were prompted by his reading of Wittgenstein. I argue that this distinctive approach allows for the identification of an original position held by Blumenberg with regard to mathematics and its practices. This position emerges not only from an analysis of how Blumenberg interprets Wittgenstein's views, but also from an examination of Blumenberg's conception of mathematical proofs.

This work was supported by a Marbach Fellowship at the German Literature Archive (DLA) in Marbach. I thank warmly Dorit Krusche for her help and advice and Juliet Floyd for her careful reading and comments. I also warmly thank Bettina Blumenberg and the German Literature Archive (DLA) in Marbach for their kind permission to cite the texts from the various cards and books from Blumenberg's Nachlass and to present various photos of them.

M. Friedman (✉)
Hausdorff School for Mathematics, University of Bonn, Bonn, Germany
e-mail: friedman@math.uni-bonn.de

A. Fragio et al. (eds.), *Hans Blumenberg's History and Philosophy of Science*, Studies in History and Philosophy of Science 63,
https://doi.org/10.1007/978-3-031-78251-0_3

1 Introduction

In the first pages of the chapter "Im Fliegenglas" from his book *Höhlenausgänge*, Hans Blumenberg stresses "the determination of the impossibility of dwelling in the execution of logic" ["die Bestimmung der Unmöglichkeit des Verweilens im Vollzug der Logik"] (1989, p. 757). Examining Wittgenstein's conception of logic, Blumenberg describes this impossibility—expressed in a certain letter that Wittgenstein wrote to Russell in 1913, which I will examine below—as a dark path. Such a darkness is caused by the execution of a complete logical investigation, and points towards the possibility of being captive in logic. One can, as Blumenberg claims regarding this possible captivity, either return to a more luminous area or continue walking along this path. Blumenberg then notes that Wittgenstein, years after writing this letter, in his *Philosophical Investigations*, underlines the relationship of every philosophy to this position of imprisonment. He cites Wittgenstein's famous metaphorical remark concerning the relationship of philosophy to this starting position: "What is your aim in Philosophy?—To show the fly the way out of the fly-bottle" (Wittgenstein, 1963, p. 407, quoted in Blumenberg, 1989, p. 757). While one may read Blumenberg's discussion as an attempt to examine Wittgenstein's metaphors—such as those of the path, light or prison—what is surprising in "Im Fliegenglas" is the example that Blumenberg brings afterward from Wittgenstein's *Schriften 6. Bemerkungen über die Grundlagen der Mathematik* (from now on BGM; Blumenberg, 1974). This example involves a geometrical puzzle illustrating how one can construct a rectangle from four, not necessarily identical, right-angled triangles, a construction which leads to a surprise since the result "is beyond the scope of the parts' possible positioning, their guidance or the seduction of optical presumptions" ["Das Ergebnis liegt außerhalb des Umkreises sich anbietender Lagerungen der Teile, ihrer Führung oder Verführung zu optischen Präsumtionen"] (Blumenberg, 1989, p. 758).

Given that Blumenberg, in his published writings, hardly addressed the philosophy or the history of mathematics as such, this example is indeed somewhat surprising.[1] In his early work, *Paradigms for a Metaphorology* from 1960, one finds two chapters dealing directly with mathematical concepts or with metaphors coming from mathematics: Chapter VIII analyzes the emergence of the mathematical concept of probability, while Chapter X considers the transition from geometrical symbols to metaphors, focusing on the circle and the sphere. However, these examples primarily serve to discuss the nature of metaphors rather than to establish a unique position in the history of mathematics. Moreover, starting in the late 1970s the anthropological aspects in Blumenberg's works take center stage, while his

[1] To stress: Blumenberg did acknowledge the role mathematics plays in the service either of natural philosophy during the early modern age or of the modern natural sciences, and in fact discussed this role in his writings, though not extensively. A thorough discussion of Blumenberg's conception of the history and philosophy of mathematics as well as of mathematical concepts is outside the scope of this paper. See, however, Friedman (2024).

studies in the history of metaphors and of the natural sciences are moved to the background. During these years, one may note, at least when inspecting the published writings, that references to mathematics appear less often. In addition, in his notes in "Theory of Nonconceptuality" from ca. 1975 Blumenberg underlines that "the desire for linguistic definiteness [...] is bound to the ideal of mathematics to cover, with the use of a constructive set of instruments, all noncontradictory possibilities" (2020, p. 282; 2007, p. 51); and even before noting this 'ideal,' he stresses that "the irresistible model of mathematics has not so much corrupted philosophical morals [...] as cut off the air for philosophical thought to breathe" (2020, p. 280; 2007, p. 49).[2]

However, reading Wittgenstein's BGM in 1984, Blumenberg notices a way in which mathematics, or more precisely, mathematical practices, do give air to philosophical thought. His remarks and insights— inspired and resulted from this reading[3]—seen, for example, in the *Höhlenausgänge* but also in his unpublished writings, can be considered as focusing on mathematical practices, and not necessarily on a critique of mathematical concepts or foundations, although he also criticizes such foundational thinking, as we will see below. Indeed, in the chapter "Im Fliegenglas" Blumenberg refers several times to Wittgenstein's views on mathematics, logic, and the mathematical proof. These references reveal an original position held by Blumenberg concerning mathematics and its practices, considering not only how Blumenberg views Wittgenstein's metaphors, but also how Blumenberg thinks about mathematical proofs. Hence, I will examine in this chapter Blumenberg's reflections on Wittgenstein's philosophy of mathematics as mainly found in Wittgenstein's BGM, which Blumenberg read, commented on, and cited. Blumenberg's exemplar of BGM, being the edition from 1974,[4] found in the German Literature Archive in Marbach, Germany, will serve as the focus of this chapter along with his markings and underlining in it, and the ways he integrated and reconceptualized Wittgenstein's reflections in his own thought. This chapter will therefore aim to unfold this unique focus in Blumenberg's thought, which till now remained unnoticed in the research on Blumenberg and his conceptualization of the history and philosophy of science.

The chapter is divided into four parts, each analyzing Blumenberg's reflections on mathematics and mathematical concepts—either published or unpublished—due

[2] "Die Sehnsucht nach sprachlicher Eindeutigkeit [...] ist [...] an das Ideal der Mathematik gebunden, durch ein konstruktives Instrumentarium alle widerspruchsfreien Möglichkeiten [...] abzudecken" and "Das bestechende Vorbild der Mathematik hat hier immer wieder zwar nicht die philosophischen Sitten verdorben [...] als vielmehr dem philosophischen Denken die Atemluft genommen." About Blumenberg's analysis in the late 1970s of the mathematical formula as such a possible model, see Hertfelder (2022).

[3] It is important to stress that Blumenberg refers to Wittgenstein not only in *Höhlenausgänge* but also in other works. He discusses Wittgenstein and his writings in *Die Sorge geht über den Fluss* (Blumenberg, 1987), in his reflections on nonconceptuality (see footnote 10 below), in *Die Vollzähligkeit der Sterne* (1997a), and in *Die Verführbarkeit des Philosophen* (2000).

[4] All references in this chapter refer to Blumenberg's exemplars of Wittgenstein's writings, found in the DLA Marbach (Signatur: G: Blumenberg, Hans (Philosophisch-literarische Teilbibliothek)).

to his reading of Wittgenstein. Section 2 begins with describing when and how Blumenberg engaged with Wittgenstein's writings, shedding light on his practices of reading and writing. Section 3 moves on to look at the geometrical puzzle that both Wittgenstein and Blumenberg reflected upon. Section 4 continues the analysis of "Im Fliegenglas," focusing on how Blumenberg considered Wittgenstein's views on the 'paradise' of mathematics: set theory, being instead considered by Wittgenstein as a parody. Section 5 turns to the unpublished writings of Blumenberg: in one of his cards, reacting to Wittgenstein's claim that 'Der Mathematiker erzeugt *Wesen*,' Blumenberg unfolds his own conception of 'secondary Platonism' concerning mathematics. In the conclusion (Sect. 6) I examine why Blumenberg may be considered as an innovative reader of Wittgenstein, implicitly offering new perspectives in the philosophy of mathematics.

2 Blumenberg Reading Wittgenstein's *Schriften*

Before examining Blumenberg's discussion of the geometrical puzzle presented in BGM, I want first to explore the practices of Blumenberg when reading and writing about Wittgenstein's writings. It is essential to note that Blumenberg thoroughly read all the writings by Wittgenstein, mostly during the 1980s. The German Literature Archive in Marbach, Germany, has Blumenberg's exemplars of all of eight Wittgenstein's *Schriften*, together with the cards on which Blumenberg wrote down the main subjects he found interesting and the dates of his reading. Although many of Wittgenstein's writings contain comments on mathematics, which Blumenberg sometimes underlined and commented upon in his cards, the examination of all of them is beyond the scope of this chapter. This chapter therefore focuses on Blumenberg's reading of BGM. It is documented that Blumenberg read the first 50 pages by 20 January 1984, and then, during September 1984, he read the rest of the book.[5] Blumenberg's method of reading BGM was more or less identical for all of Wittgenstein's writings: while reading such a volume, he underlined chosen sentences; once he found the topic interesting, he wrote it and the corresponding page number on a card (see Fig. 1). Moreover, Blumenberg's own reflections on specific statements of Wittgenstein, reflections which were written either during the reading or after the reading was completed (when going over the card with the topics), were usually written on a separate card. If this was the case, a small letter 'k' (from 'Karte,' i.e. card) was also written next to the corresponding section (see Fig. 2). While several of these cards were later integrated into the published writings of Blumenberg, this is not necessarily true for all the cards Blumenberg wrote.

Blumenberg's own documentation shows that he had several periods of concentrated reading of Wittgenstein's writings. According to his reading lists, Blumenberg

[5] See Blumenberg, Hans: Einlagen 225 (of Wittgenstein's *Schriften 6. Bemerkungen über die Grundlagen der Mathematik*), DLA Marbach, Bestand C: Blumenberg, Hans.

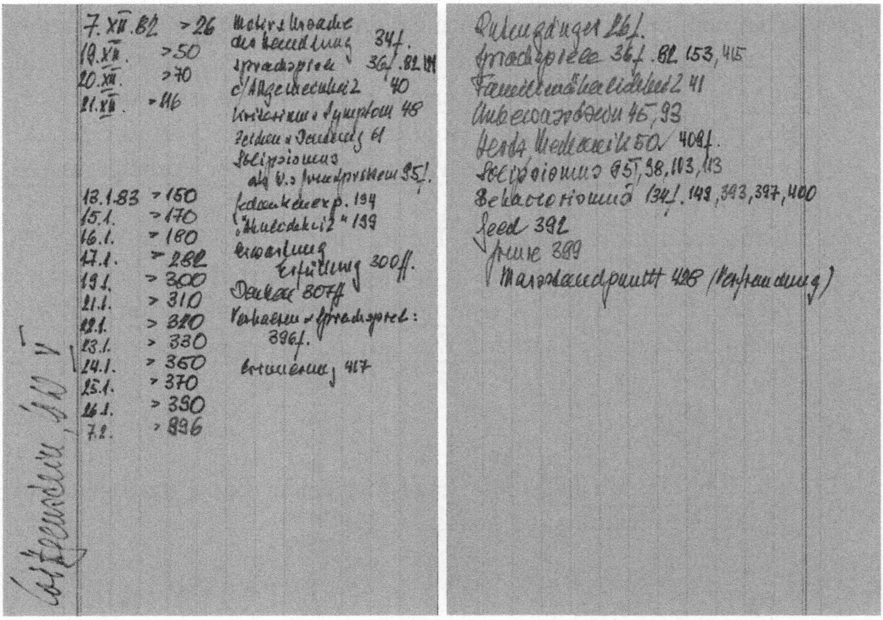

Fig. 1 The recto and verso side of Blumenberg, Hans: Einlagen 224 (of Wittgenstein's *Schriften 5. Das blaue Buch* (DLA Marbach, Bestand C: Blumenberg, Hans), indicating the dates of reading and selected topics along the corresponding page numbers

first read Wittgenstein's *Tractatus* in 1965. One can find in the reading list ('Leseliste') of the years 1960–1972 the following entry under the year 1965: "3.VIII Wittgenstein: Tractatus logico-philosophicus," which means probably that his reading was completed on 3 August 1965.[6] In 1977, Blumenberg read Wittgenstein's *Vermischte Bemerkungen*, containing a collection of notes from Wittgenstein's estate.[7] A more concentrated reading took place during the 1980s. According to his cards, between 18.10.1982 and 25.2.1983 Blumenberg read, in the following order, *Schriften* (subtitled: *Tractatus logico-philosophicus. Tagebücher 1914–1916. Philosophische Untersuchungen*), *Schriften* 2 (*Philosophische Bemerkungen*), *Schriften 5* (*Das blaue Buch. Eine philosophische Betrachtung. Zettel*), and *Schriften 3* (*Wittgenstein und der Wiener Kreis*). Moreover, Blumenberg may have reread parts of the *Tractatus* around 1982, which he had read already in 1965, as noted above. Indeed, one can find on page 83 of Wittgenstein's *Schriften*, being the last page of the *Tractatus*, two additions in Blumenberg's handwriting with different types of ink (see Fig. 3). Both refer to the ladder that should be thrown

[6] In Blumenberg, Hans: Leseliste 1960–1972, DLA Marbach, Bestand A: Blumenberg, Hans.

[7] This is, according to the reading list for the years 1973–1986, under the year 1977, where one finds the entry "5.X. Wittgenstein: vermischte Bemerkungen." Here and in the following paragraphs, I base my findings on Blumenberg, Hans: Leseliste 1973–1986, DLA Marbach, Bestand A: Blumenberg, Hans.

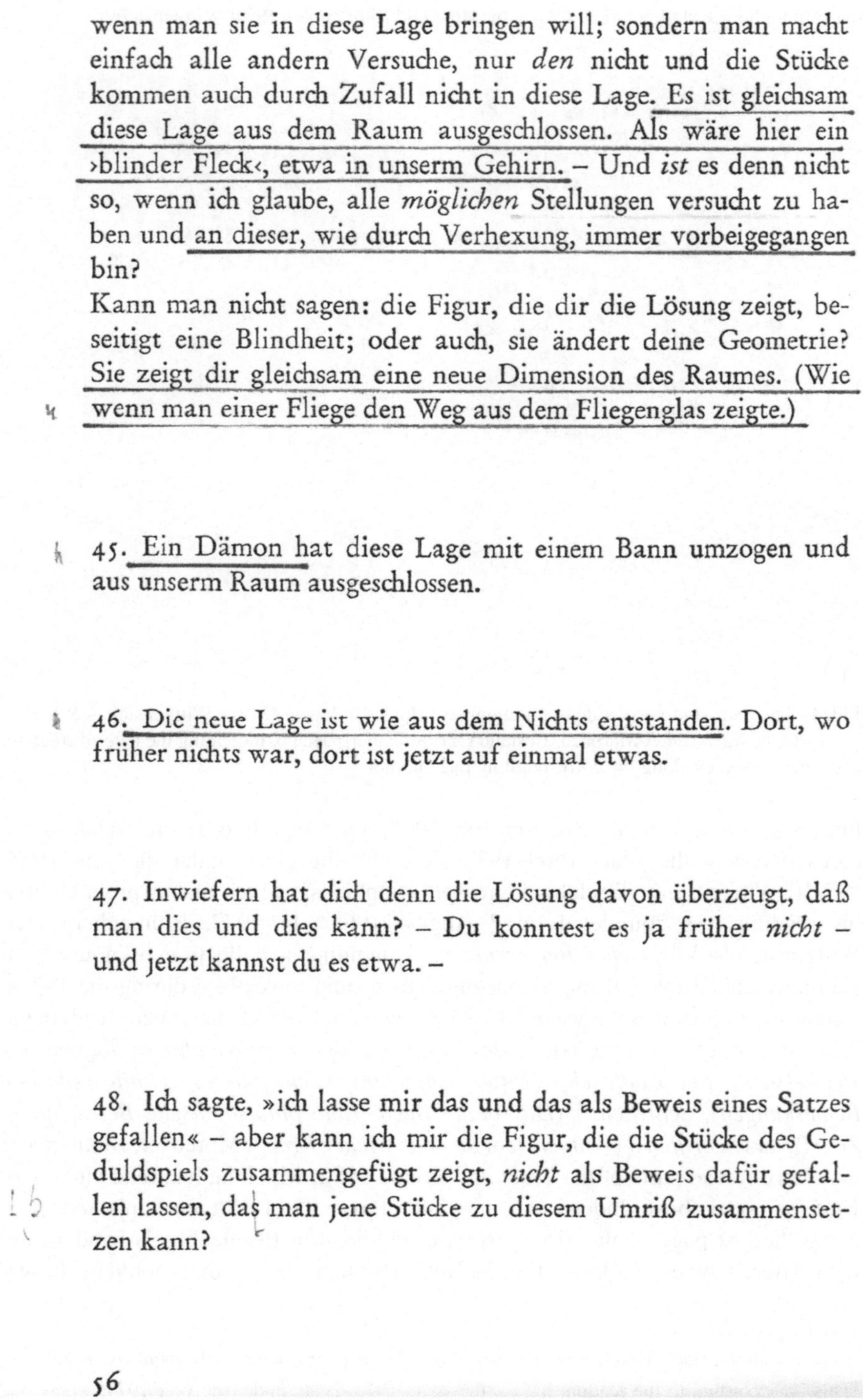
wenn man sie in diese Lage bringen will; sondern man macht einfach alle andern Versuche, nur *den* nicht und die Stücke kommen auch durch Zufall nicht in diese Lage. Es ist gleichsam diese Lage aus dem Raum ausgeschlossen. Als wäre hier ein ›blinder Fleck‹, etwa in unserm Gehirn. – Und *ist* es denn nicht so, wenn ich glaube, alle *möglichen* Stellungen versucht zu haben und an dieser, wie durch Verhexung, immer vorbeigegangen bin?
Kann man nicht sagen: die Figur, die dir die Lösung zeigt, beseitigt eine Blindheit; oder auch, sie ändert deine Geometrie? Sie zeigt dir gleichsam eine neue Dimension des Raumes. (Wie wenn man einer Fliege den Weg aus dem Fliegenglas zeigte.)

45. Ein Dämon hat diese Lage mit einem Bann umzogen und aus unserm Raum ausgeschlossen.

46. Die neue Lage ist wie aus dem Nichts entstanden. Dort, wo früher nichts war, dort ist jetzt auf einmal etwas.

47. Inwiefern hat dich denn die Lösung davon überzeugt, daß man dies und dies kann? – Du konntest es ja früher *nicht* – und jetzt kannst du es etwa. –

48. Ich sagte, »ich lasse mir das und das als Beweis eines Satzes gefallen« – aber kann ich mir die Figur, die die Stücke des Geduldspiels zusammengefügt zeigt, *nicht* als Beweis dafür gefallen lassen, daß man jene Stücke zu diesem Umriß zusammensetzen kann?

56

Fig. 2 Page 56 of Blumenberg's exemplar of Wittgenstein's BGM, where several sentences are underlined and marked with the letter 'k,' and (on the next page) Blumenberg's corresponding card to section 45 (the recto side of Blumenberg, Hans: POS WTG: Sein Begriff vom "Raum" & der Mythos vom Dämon (Nr. 024345), DLA Marbach, Bestand A: Blumenberg, Hans, Zettelkasten, 11: P–S)

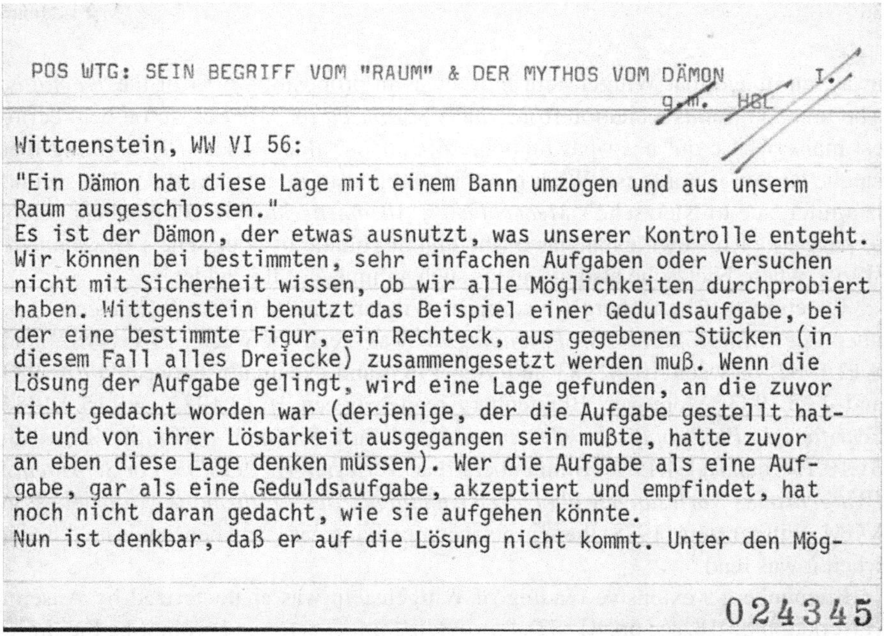

POS WTG: SEIN BEGRIFF VOM "RAUM" & DER MYTHOS VOM DÄMON I.
g.m. HBL

Wittgenstein, WW VI 56:
"Ein Dämon hat diese Lage mit einem Bann umzogen und aus unserm Raum ausgeschlossen."
Es ist der Dämon, der etwas ausnutzt, was unserer Kontrolle entgeht. Wir können bei bestimmten, sehr einfachen Aufgaben oder Versuchen nicht mit Sicherheit wissen, ob wir alle Möglichkeiten durchprobiert haben. Wittgenstein benutzt das Beispiel einer Geduldsaufgabe, bei der eine bestimmte Figur, ein Rechteck, aus gegebenen Stücken (in diesem Fall alles Dreiecke) zusammengesetzt werden muß. Wenn die Lösung der Aufgabe gelingt, wird eine Lage gefunden, an die zuvor nicht gedacht worden war (derjenige, der die Aufgabe gestellt hatte und von ihrer Lösbarkeit ausgegangen sein mußte, hatte zuvor an eben diese Lage denken müssen). Wer die Aufgabe als eine Aufgabe, gar als eine Geduldsaufgabe, akzeptiert und empfindet, hat noch nicht daran gedacht, wie sie aufgehen könnte.
Nun ist denkbar, daß er auf die Lösung nicht kommt. Unter den Mög-

024345

Fig. 2 (continued)

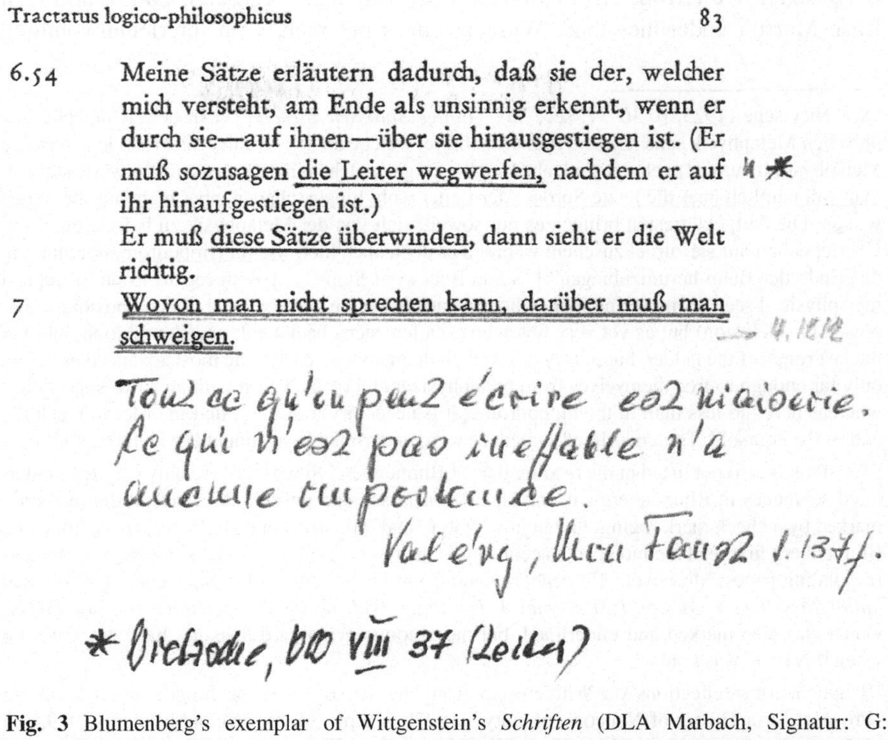

Tractatus logico-philosophicus 83

6.54 Meine Sätze erläutern dadurch, daß sie der, welcher mich versteht, am Ende als unsinnig erkennt, wenn er durch sie—auf ihnen—über sie hinausgestiegen ist. (Er muß sozusagen die Leiter wegwerfen, nachdem er auf ihr hinaufgestiegen ist.) M *
Er muß diese Sätze überwinden, dann sieht er die Welt richtig.

7 Wovon man nicht sprechen kann, darüber muß man schweigen. → 4.12.12

Tout ce qu'on peut écrire est médiocre.
Ce qui n'est pas ineffable n'a aucune importance.
Valéry, Mon Faust I.137f.

* Vorrede, MA VIII 37 (Leiter)

Fig. 3 Blumenberg's exemplar of Wittgenstein's *Schriften* (DLA Marbach, Signatur: G: Blumenberg, Hans (Philosophisch-literarische Teilbibliothek), p. 83, being the last page of the *Tractatus*, in which Blumenberg added two notes, probably in two different times: one a citation from Valéry's *Mon Faust*, and the other a reference to page 37 of Nietzsche's *Menschliches, Allzumenschliches*

away, an image that Wittgenstein introduces in proposition 6.54 of the *Tractatus*. The first addition is a citation from Valéry's *Mon Faust*: 'Tout ce qu'on peut écrire est niaiserie. Ce qui n'est pas ineffable n'a aucune importance' ['Everything that can be written is nonsense. What is not ineffable is of no importance']. The second is a reference to Nietzsche's *Menschliches, Allzumenschliches*. Blumenberg refers to page 37 of his own exemplar of the eighth volume of Nietzsche's *Gesammelte Werke*, where Nietzsche also discusses such an image of the ladder.[8]

To return to Blumenberg's reading of Wittgenstein, *Schriften 8* (*Bemerkungen über die Philosophie der Psychologie*) was read between 19.11.1982 and 8.12.1982, between 10.4.1983 and 16.4.1983, and eventually between 17.8.1983 and 27.8.1983. Moreover, Blumenberg read between 16.12.1983 and 15.9.1984 *Schriften 4* (*Philosophische Grammatik*) and then *Schriften 6* (BGM). Between 31.10.1985 and 11.4.1986, Blumenberg read *Vorlesungen* 1930/35. For *Schriften 7* (*Wittgensteins Vorlesungen über die Grundlagen der Mathematik* (from now on VGM; Wittgenstein, 1978) there is no accompanying card, and hence it is not known when it was read.[9]

Blumenberg's extensive reading of Wittgenstein was characterized by Anselm Haverkamp (2018, see pp. 81–82) as providing impetus for and prompting the 'turn' from metaphorology to his reflections on the theory of nonconceptuality.[10] However, the secondary literature on the relations between the Blumenberg and Wittgenstein is not extensive (Kroß, 2010; Marrou, 2016; Menke, 2009; Zill, 2008), although Élise Marrou underlines that "Wittgenstein is not merely an interlocutor among

---

[8] See Nietzsche (1923, p. 36–37, sec. 20): "*Einige Sprossen zurück* [...] In betreff der philosophischen Metaphysik sehe ich jetzt immer mehrere, welche an das negative Ziel (daß jede positive Metaphysik Irrtum ist) gelangt sind, aber noch wenige, welche einige Sprossen rückwärts steigen; man soll nämlich über die letzte Sprosse der Leiter wohl hinausschauen, aber nicht auf ihr stehen wollen. Die Aufgeklärtesten bringen es nur soweit, sich von der Metaphysik zu befreien und mit Überlegenheit auf sie zurückzusehen: während es doch auch hier, wie im Hippodrom, not thut, um das Ende der Bahn herumzubiegen" ["Some Backward Steps [...] With regard to philosophical metaphysics I see ever more and more who have arrived at the negative goal (that all positive metaphysic is a delusion) but as yet very few who go a few steps backward: one should look out over the last rungs of the ladder, but not try to stand on them, that is to say. The most advanced as yet go only far enough to free themselves from metaphysic and look back at it with an air of superiority: whereas here, no less than in the hippodrome, it is necessary to turn around in order to reach the end of the course"]. The underlined sentences were underlined by Blumenberg in his exemplar.

[9] The book is also not listed in the reading lists of Blumenberg. Since there are only very few underlined sentences in Blumenberg's exemplar of *Schriften 7* and only three entries at the index are marked by a check mark (being 'Gefängnis,' 'Gott,' and 'Hilbert'), one might perhaps assume that Blumenberg first read the index and, according to the themes he found interesting, he read the corresponding pages. Moreover, Blumenberg had also in his possession Wittgenstein's *Briefwechsel mit B. Russell, G. E. Moore, J. M. Keynes, F. P. Ramsey, W. Eccles, P. Engelmann und L. von Ficker*, which was also marked and underlined, but the accompanying card does not have any dates on when this book was read.

[10] Blumenberg's reflections on Wittgenstein from the 1970s are to be found, for example, in "Prospect for a Theory of Nonconceptuality" (1997b, see pp. 84, 89–90), published in 1979, and in the earlier notes from ca. 1975, published as "Theorie der Unbegrifflichkeit" (2007, see pp. 37–38, 99, 102–103).

others for Blumenberg, but one of his doubles" (2016, p. 109). Marrou (2013) also stresses that Blumenberg does not distinguish between the early and the later Wittgenstein, nor between biographical episodes and philosophical writings or remarks. Moreover, Blumenberg's reconceptualization of Wittgenstein's views on mathematics in "Im Fliegenglas" remains almost unnoticed in the literature.

Indeed, Wittgenstein's BGM and his other writings address a variety of mathematical subjects. While the *Tractatus* is definitely considered an exemplary work representing a highly succinct form of logical analysis, other works deal with different areas and practices of mathematics. Although this chapter does not aim to review all of these works, one subject discussed in Wittgenstein's writings is worth mentioning: the mathematical foundational crisis, or as is better known, the *Grundlagenkrise*.

Here, a brief detour is needed to explain the nature of this crisis. This crisis,[11] one may say, had its roots in the assumption of naïve set theory: given any well-defined logical or mathematical property, there exists a corresponding well-defined set of all objects that satisfy that property. For example, the property 'being an orange' corresponds to the set of all the oranges. This assumption led, starting at the end of the nineteenth century, to the discovery of several paradoxes. One of the most famous paradoxes, published in 1901, which arises from this naïve assumption, is the well-known Russell paradox, based on the property: x does not belong to itself.[12] Wittgenstein discusses this paradox extensively, and Blumenberg was certainly aware of this discussion, as well as of the paradox itself (Wittgenstein, 1978, see p. 270ff.).[13] These paradoxes, among others, led to questions about the consistency of set theory and arithmetic, eventually culminating in the foundational crisis of mathematics. They prompted as well the development of a more rigorous axiomatic basis of set theory: either with Russell's type theory or with the formalization of the axioms of ZFC (which stands for 'Zermelo-Fraenkel axioms with the axiom of choice'). What followed was the clash between two (philosophical) schools of mathematics: between the "method" of formalism and the "full blown philosophy" of intuitionism (Ferreirós, 2007, p. 332). The formalism's method reached its peak with David Hilbert's program, which attempted, during the 1920s, to ground various domains of mathematics on a finite, consistent, and complete set of axioms. However, such a program met its end with the publication of Gödel's incompleteness theorems in 1931 (see also below). Having read Wittgenstein's writings meticulously, it is reasonable to suppose that Blumenberg was aware at least of the main

[11] This foundational crisis is composed of several episodes, and the following review is only a rough summary. For a full survey, see, e.g., Ferreirós (2007, see pp. 297–336, and 2008).

[12] Based on this property one constructs the set: $R = \{x : x \notin x\}$, where $\in$ denotes the property of belonging. This construction leads to a contradiction: if $R \in R$, then $R \notin R$ (according to the definition of R), and similarly, if $R \notin R$, then $R \in R$. A well-known Russell-like paradox is the liar paradox or the barber paradox.

[13] In a card titled "Russels Antinomie d Mengenlehre" Blumenberg describes Russell's paradox. The card is dated both April 1979 and May 1983 (see Blumenberg, Hans: Russels Antinomie d Mengenlehre (Nr. 7453), DLA Marbach, Bestand A: Blumenberg, Hans, Zettelkasten 07 – K–M – Logik).

episodes of this crisis. Indeed, this crisis is discussed, for example, in several sections of VGM (for example, on page 13f. or page 272), which Blumenberg read and underlined in part.[14]

3 Spaces of Mathematical Practice I: Blumenberg on Wittgenstein's Geometrical Riddle

Having explained how Blumenberg read Wittgenstein's writings, I turn to discuss the chapter "Im Fliegenglas." Blumenberg begins this chapter with a citation from a letter that Wittgenstein wrote to Bertrand Russell on March 25, 1913. In this letter, Wittgenstein reflects that whenever he thinks about logic, his thoughts become so vague "as if one was led by someone through a dark corridor with a light, and just when one gets to the middle, the light goes out and one is left alone" ["als ob man von jemandem mit einem Licht durch einen dunklen Gang geführt würde, und gerade wenn man in der Mitte ankommt, geht das Licht aus, und man ist allein gelassen"] (Blumenberg, 1989, p. 753). Blumenberg first analyzes this statement, which employs a metaphor of extinguishing and extinguished light, with a reference to Otto Neurath's lecture "The Lost Wanderers of Descartes and the Auxiliary Motive" ("Die Verirrten des Cartesius und das Auxiliarmotiv") from 1913. However, Blumenberg stresses that it is not clear whether Wittgenstein heard this lecture or knew of it. More central to our discussion is not Blumenberg's reading of Neurath, but how he emphasizes another aspect in Wittgenstein's citation, which is very much reflected, so Blumenberg, in the later writings of Wittgenstein.

For Blumenberg, it is logic that leaves the wanderer stranded in the mathematical cave: "its light is not enough for the whole course, due to the immensity of darkness" ["Ihr Licht reicht nicht für den ganzen Gang, für das Unmaß an Dunkelheit"] (1989, p. 756). Blumenberg's references to the extinguishing light of logic are implicitly explained when he notes how Wittgenstein dealt with the "elimination of the question of the provable freedom from contradiction of logical-mathematical propositional systems" ["Ausschaltung der Frage nach der beweisbaren Widerspruchsfreiheit logisch-mathematischer Aussagensysteme"] (1989, p. 755). This reference is here to the attempts done during the first decades of the twentieth century to prove the consistency of several mathematical domains and their

---

[14] For example, in Blumenberg's exemplar of VGM, the following sentence at page 272 is underlined: "Denken wir an Russells Widerspruch und nehmen wir an, man hätte ihn niemals gefunden. Sollen wir deshalb sagen, die Grundlagen der Mathematik seien falsch gewesen?" ["Consider Russell's contradiction, and suppose that it had never been found. Should we say that on account of this, the foundations of mathematics would have been wrong?"]. In addition, one of the articles in Blumenberg's *Konvolut: Materialsammlung Mathematik/Physik* (DLA Marbach, Bestand: A: Blumenberg, Hans) is that of Friedrich Kambartel: "Philosophische Perspektiven der Diskussion um die Grundlagen der Mathematik" from 1963. Kambartel (1963, see pp. 159–160, 190ff.) discusses there extensively the foundational crisis.

associated axiomatic systems: logic, arithmetic, geometry, and, in general, mathematics. These attempts, to recall the short review above concerning the foundational crisis, ended up in a failure with the proof of Gödel's incompleteness theorems, and particularly with his second theorem, which states that the consistency of any system, which can express the basic arithmetic of natural numbers, cannot be proved in this system itself. While Blumenberg is aware of this episode,[15] he does not discuss it—at least not directly. However, as we will see, he does discuss Wittgenstein's reading of the consequences of the foundational crisis. Blumenberg observes that for the later Wittgenstein this parable on the extinguishing light of mathematics in the cave shows that one must nevertheless continue "on a path even when the warnings of paradoxes, [and] the threats of contradictions have become unmistakable" ["auf einem Weg auch dann weiterzugehen, wenn die Warnungen vor Paradoxien, die Drohungen von Widersprüchen unüberhörbar geworden sind"]. Hence, one should also continue when the "logical orientation" fails (1989, p. 756). Such logical orientation and, generally, the logical tools employed are not an "autonomous instrument, to find one's way in the world" (Blumenberg, 1989, p. 756). In this sense, Blumenberg aligns with the critique of Husserl on mathematics: while mathematics was thought during the early modern age to reach the "naked truth" (1981, p. 30), "what looked like nakedness turns out to be a 'well-fitting garb of ideas' whose measurements are taken in the geometrical and natural-scientific mathematization of the lifeworld—so Edmund Husserl in his interpretation of Galileo" (2010, p. 48). In fact, Blumenberg suggests the opposite—wandering in the mathematical 'caves,' while, for example, trying to find a proof, is as if one walks in an underground cave, on a dark path ["dunkler Gang"] (1989, p. 757), where the logical 'instrument' does not provide any guarantee to solve problems.

At this stage Blumenberg turns to deal with concrete mathematical problems, or more precisely, riddles. Blumenberg cites explicitly from BGM, discussing what happens when one attempts to solve "mathematical riddle questions" ["mathematische Rätselfragen"] (1989, p. 757). Wittgenstein discusses this issue in *Anhang II* to Part I of BGM (1974, see p. 111), when he asks in what sense, and if at all, something is surprising in mathematics. Two reasons for a possible surprise in mathematics are presented: first, that the mathematical series of thought ["Gedankenreihe"] allows something surprising to emerge; second, that it is surprising in which "depths the mathematical investigation penetrates" ["in welche Tiefe die mathematische Untersuchung dringt"] (Wittgenstein, 1974, p. 111), a sentence

[15] See above; also note that in Blumenberg's exemplar of BGM the following sentence is underlined in p. 386: "Gödel's proposition, which asserts something about itself, does not *mention* itself" ["Der Gödelsche Satz, der etwas über sich selbst aussagt, *erwähnt* sich selbst nicht"]. Other sections in BGM, which deal with Gödel's theorems, are also underlined by Blumenberg (see, e.g., sentences in p. 120, sections 12 and 13; or p. 122, section 19; or p. 383 and 388). Kambartel (1963, see pp. 168–169) discusses Gödel's theorems, and Wittgenstein is also discussed in Kambartel's paper (1963, see p. 183ff.), one which Blumenberg read. Moreover, Blumenberg suggested in the late 1960s to publish Kambartel's Habilitation *Erfahrung und Struktur* (Zill, 2020, see p. 296), which was indeed published in 1968. Gödel's theorems are also mentioned there as well as Wittgenstein (Kambartel, 1968).

which is underlined by Blumenberg in his copy of BGM. In the next section Wittgenstein stresses a claim, which is also underlined by Blumenberg, that "the mathematician is an inventor, not a discoverer" (1974, p. 111).[16] Moreover, a surprise may be not legitimate, since the moment we understand the proof, "the surprise vanishes" (Wittgenstein, 1974, p. 112). Here Wittgenstein employs the metaphor of a path leading to a mathematical proof in a sentence which is also underlined by Blumenberg: "That this path, which is all before me, [...] that is no surprise" ["Dass dieser Weg, den ich ganz von mir habe, [...] das ist keine Überraschung"] (Wittgenstein, 1974, p. 112). In the next section the following sentences are also underlined in Blumenberg's exemplar of BGM: Wittgenstein discusses there a solution to a mathematical riddle, and the impossibility of understanding it at first: "as if we had walked through an underground passage and were now coming to light somewhere" ["als seien wir durch einen unterirdischen Gang gegangen und kämen nun irgendwo ans Licht"] (Wittgenstein, 1974, p. 112). These sentences are also cited in *Höhlenausgänge* (Blumenberg, 1989, p. 757), and Blumenberg, in addition, quotes Wittgenstein's claim that this found solution is equivalent to not knowing where we are, and also not knowing where the tunnel's entrance or exit is.

The question 'what an underground movement is concerning solutions to mathematical problems,' may be answered with the example that Blumenberg brings from Wittgenstein's BGM. This example is not drawn from *Anhang II*, but rather from earlier sections of Part I of BGM. In sections 42–50, Wittgenstein discusses mathematical discovery with two riddles: the first one instructs to compose a rectangle from four (non-identical) right-angled triangles; the second composes a rectangle from two parallelograms and two triangles. Both riddles are accompanied by diagrams, whereas the diagram for the second riddle is called by Wittgenstein a "proof" (1974, p. 57, section 50; see Fig. 4). However, none of these diagrams are reproduced by Blumenberg.

Wittgenstein stresses already for the first riddle that someone who sees its solution finds that they did not think about it earlier: "The new position [of the shapes] has as it were come to be out of nothingness" ["Die neue Lage ist wie aus dem Nichts entstanden. Dort, wo früher nichts war, dort ist jetzt auf einmal etwas"] (1974, p. 56). He observes concerning this position of the triangles that "a demon has cast a spell round this position and excluded it from our space" (1974, p. 56). This is the reason why we can neither conceive nor imagine this position. In addition, when we see it, it is as if it shows us "a new dimension of space. (As if a fly were shown the way out of the fly-bottle)" (1974, p. 56). Here, an explicit example of how such a fly can find its way out is provided: the 'aim of philosophy,' to refer to the citation from *Philosophical Investigations*, is explicated via a mathematical example.

[16] That this philosophical claim was of certain importance to Blumenberg is to be seen in the fact that in his insert to Wittgenstein's BGM, he writes down: "Erfinden oder Entdecker 99, 136, 166" (Blumenberg, Hans: Einlagen 225, DLA Marbach, Bestand C: Blumenberg, Hans).

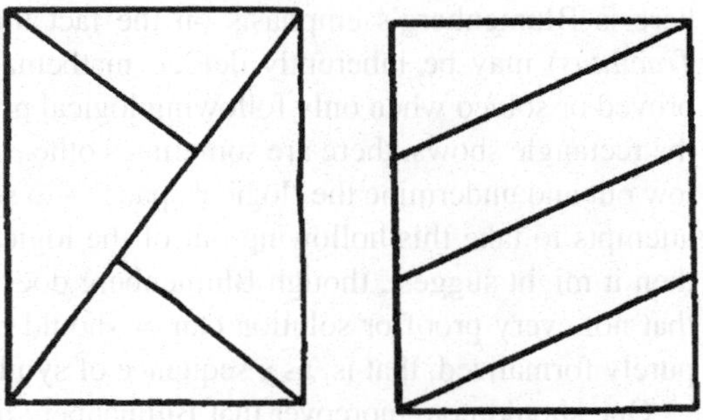

Fig. 4 Wittgenstein's diagrams for two of the riddles he presents in BGM (Wittgenstein, 1974, pp. 55, 57). Blumenberg refers to the left diagram in the *Höhlenausgänge*

Presenting the later Wittgenstein as someone who developed *avant la lettre* a philosophy of mathematical diagrams and practices, one can argue that a diagram can also function as a proof. It demonstrates how mathematical proofs are practiced, underlining that diagrams and visual reasoning can be an integral part of the mathematical proof.[17] Blumenberg's focus is slightly different, and he is more interested in the metaphors Wittgenstein employs. Blumenberg underlines the above-cited sentences in his exemplar of BGM, and they are also cited in *Höhlenausgänge*. It is clear why these sections are of interest to Blumenberg: the figure of the fly as well as of the demon are to be found also in Blumenberg's other writings (Nicholls, 2014), and he returns to the demon in "Im Fliegenglas," as we will see below.[18] Blumenberg stresses that the demon is neither a scammer nor a deceiver, just as the diagram, showing us the solution, has not "hidden or falsified" anything (1989, p. 759), since there was *nothing*, so Blumenberg, neither in the riddle nor in its solutions, to hide. Nevertheless, while Wittgenstein notes that such solutions show a 'new dimension' of space, in the sense that they offer a new way to look at space, or—more precisely—at geometrical figures and practices, Blumenberg stresses another aspect concerning Wittgenstein's demon. Noting that demons and angels are not yet thoroughly discussed in Wittgenstein's works, Blumenberg emphasizes that "through them, intricacies, as well as licenses in thought experiments, are created, especially imaginary spaces like caverns [*Kavernen*] in the 'logical space' of the *Tractatus*" ["Durch sie werden Vertracktheiten wie Lizenzen in Gedankenexperimenten erzeugt, vor allem imaginäre Raume wie Kavernen im ‚logischen Raum' des ‚Tractatus'"] (1989, p. 758, n. 75). Similar to how he stresses Wittgenstein's conception of a proof as an underground passage to where the 'light' of logic does not reach, here the 'demon,' which shows the diagrammatical proof, prompts the generation of such hollow spaces in the logical space. Essential to note

[17] Following Pérez-Escobar, "[a]ccording to the received view, the first elaborated philosophical work concerned with the practical aspects of mathematics was crafted by Lakatos" (2022, p. 162). Pérez-Escobar claims that one should also consider later Wittgenstein's philosophy as a precursor of this tradition. However, it is not clear whether Blumenberg knew of Lakatos' work on the philosophy of mathematics.

[18] In Blumenberg's insert to BGM one finds the following line: „Dämon 158f, 205!"(crossed over in the original) (Blumenberg, Hans: Einlagen 225, DLA Marbach, Bestand C: Blumenberg, Hans).

here is Blumenberg's emphasis on the fact that the 'logical space' (e.g., of the *Tractatus*) may be inherently defect: mathematical theorems and riddles are not proved or solved when only following logical procedures. In fact, as the example of the rectangle shows, there are sometimes other mathematical ways—ways that hollow out and undermine the 'logical space'—to solve mathematical problems. If one attempts to take this hollowing out of the logical space to its extreme conclusion, then it might suggest, though Blumenberg does not explicitly frame it in this way, that not every proof or solution can or should be presented from the beginning as purely formalized, that is, as a sequence of symbols.

One should note moreover that Blumenberg adopts a somewhat ambivalent position concerning the introduction of new proofs, as presented in the above examples. Discussing the notes from Wittgenstein's BGM written in 1942–1944, Blumenberg cites the opening sentence of section 37: "A human being is imprisoned in a room, if the door is unlocked but opens inwards; he, however, never gets the idea of pulling instead of pushing against it" ["Ein Mensch ist in einem Zimmer *gefangen*, wenn die Tür unversperrt ist, sich nach innen öffnet; er aber nicht auf die Idee kommt zu *ziehen*, statt gegen sie zu drücken"] (1989, p. 766). Wittgenstein examines in this section the usage of new or different techniques in proofs, noting before that the effect of a proof would be that the practitioner would burst or "plunge into" ["hineinstürzt"] a new rule (1989, p. 766), which is equivalent of using the door differently. Such a process parallels to how Wittgenstein introduces a surprising solution for the (de)composition of the rectangle. Blumenberg, however, reads this statement more carefully: on the one hand, he notes explicitly that the function of doors is to make possible the "*leaving* of spaces," allowing one to step outside the "logical space," creating room for other methods of mathematical proofs (1989, p. 769); on the other hand, Blumenberg also notes that such bursting of a new rule is to be considered with "caution against the loss of freedom of movement" (1989, p. 768), since once one leaves the logical space, one may no longer know how to proceed.

4 Spaces of Mathematical Practice II: Blumenberg on Hilbert's Paradise and Wittgenstein's Parody

Having seen Blumenberg's reflections on new or surprising proofs or solutions, the later reflections in "Im Fliegenglas" on mathematics move on to discuss mathematical practices in a more general manner. Here one finds one of Blumenberg's almost single references to David Hilbert (1862–1943), one of the most important mathematicians of the first half of the twentieth century. Blumenberg's method of integrating citations from various sources, sometimes recontextualizing them to present his own conception of the theme discussed, is exemplified also here when he discusses Wittgenstein's notes on Hilbert (Blumenberg, 1989, see pp. 789–790). Blumenberg begins with a brief discussion of Wittgenstein's *Schriften 4. Philosophische*

Grammatik, Part V, section 25, which contains reflections written in 1933–1934. Wittgenstein calls for greater sensitivity among mathematicians toward their own works and warns against standardization and formalization; this sensitivity is sometimes lacking in such works (Wittgenstein, 1969, see p. 381), and one should consider the plurality of plays, and hence of practices, of mathematics, instead of trying to reduce them to logic or to set theory. Blumenberg, instead of continuing to analyze this section further, returns to Wittgenstein's BGM, i.e. to the notes written in 1942–1944, and highlights the parodical-cynical remark of Wittgenstein there:

> Imagine set theory's having been invented by a satirist as a kind of parody of mathematics. —Later a reasonable meaning was seen in it and it was incorporated into mathematics. (For if one person can see it as a paradise of mathematicians, why should not another see it as a joke?) (1974, p. 264).[19]

As Blumenberg notes, Wittgenstein's reference here is to Hilbert's 1926 statement: "from the paradise [of set theory], that [Georg] Cantor created for us, no one shall be able to expel us" ["Aus dem Paradies, das Cantor uns geschaffen, soll uns niemand vertreiben können"] (Hilbert, 1926, p. 170).[20] This citation is not explicitly stated in BGM, but only a reference to it in a footnote is given; nevertheless, Blumenberg reads this citation in Wittgenstein's VGM, where the statement is explicitly underlined in his exemplar. The next sentence in VGM which is also underlined by Blumenberg is Wittgenstein's statement that he would try "to show that it [set theory] is not a paradise at all—so that one leaves of one's own accord" ["zu zeigen, dass es gar kein Paradies ist—so dass man aus eigenem Antriebe verlässt"] (Wittgenstein, 1978, p. 121). Wittgenstein does mention the context of this citation, being the paradoxes revealed in set theory, paradoxes that likely resonate with how Blumenberg aims to conceptualize the philosophy of mathematics (or, better to say, the philosophy of mathematical practices). However, in "Im Fliegenglas," Blumenberg does not provide any historical context. He also does not follow Wittgenstein's cynical remark that set theory is not at all a paradise; rather he aims to reflect upon the association of set theory as a space which is *between* paradise and parody.

In *Zettelkasten* 22 of Blumenberg's Nachlass, in a card titled "ZWH WTG: Mengenlehre = Paradies oder Parodie d Mathem,"[21] Blumenberg elaborates on this episode, which is indeed later introduced in "Im Fliegenglas." It shows a certain acquittance with the reasons why Hilbert expressed this statement:

---

[19] "Denke dir die Mengenlehre wäre als eine Art Parodie auf die Mathematik von einem Satiriker erfunden worden. —Später hätte man dann einen vernünftigen Sinn gesehen und sie in die Mathematik einbezogen. (Denn wenn der eine sie als das Paradies der Mathematiker ansehen kann, warum nicht ein andrer als einen Scherz?)."

[20] Georg Cantor (1845–1918) is known to be the main contributor to the foundations of set theory during the second half of the nineteenth century.

[21] Blumenberg, Hans: ZWH WTG: Mengenlehre = Paradies oder Parodie d Mathem, DLA Marbach, Bestand A: Blumenberg, Hans, Zettelkasten 22. The card is not numbered, but on the verso, one finds the following: "10/88 HGL ERW Sp. 713. A. 114."

What is typical of someone who tries to see the usual and common differently than it has been seen for a long time, and does not have the good fortune to be able to do so as a painter? 'Imagine set theory having been invented by a satirist as a kind of parody on mathematics. – Later a reasonable meaning was seen in it and it was incorporated into mathematics.' (Wittgenstein, WW VI 264) David Hilbert described set theory as a mathematician's paradise in his 1926 essay 'Über das Unendliche' in the *Mathematische Annalen* vol. 95. It is clear why he did this, and consistently so. But despite the association of the words, paradise and parody, it naturally makes sense to speak of a mathematicians' paradise, because there they are furthest removed from the sin of mere arithmetic and its continuations, but to speak of parody and to assume that it has been misjudged as such and subsequently incorporated into the ordinary stock of serious mathematical procedures presupposes the will to drastic outsideness. A satirist mocks the mathematicians and is then appointed a mathematician by them. This is reminiscent of procedures that are otherwise only possible in modern aesthetics: a product makes a mockery of any proximity to art, which is precisely why it will one day be found as a work of art in a museum. The common trick is to make the origin and intention invisible.[22]

Blumenberg notes in this card that what was perhaps considered first as a parody, a satire—in this case, set theory—is then turned into mathematics itself; hence set theory is between parody and paradise. Moreover, I claim that both this card and the chapter "Im Fliegenglas" stress two aspects that at first may seem unrelated: the "infantile return" (1989, p. 789) of mathematics to what at first seems silly (that is, set theory), and the metaphor of sunlight and the darkness of the caves. Indeed, in a particularly spiteful comment, after his plaidoyer to greater sensitivity from the side of mathematicians, Wittgenstein also goes against several philosophers of mathematics: "Philosophical clarity will have the same influence on the growth of mathematics as sunlight has on the growth of potato shoots. (They grow meters long in a dark cellar)" ["Die Philosophische Klarheit wird auf das Wachstum der Mathematik den gleichen Einfluß haben, wie das Sonnenlicht auf das Wachsen der Kartoffeltriebe. (Im dunklen Keller wachsen sie meterlang)"] (1969, p. 381).[23] Comparing, a

[22] "Was ist typisch für den, der versucht, das Uebliche und Gängige anders zu sehen als es seit langem gesehen worden ist, and nicht das Glück hat, dies als Maler tun zu können? "Denke dir die Mengenlehre wäre als eine Art Parodie auf die Mathematik von einem Satiriker erfunden worden. —Später hätte man dann einen vernünftigen Sinn gesehen und sie in die Mathematik einbezogen." (Wittgenstein, WW VI 264) David Hilbert hatte die Mengenlehre 1926 in einem Aufsatz "Über das Unendliche" in den "Mathematischen Annalen" Bd. 95 als das Paradies der Mathematiker bezeichnet. Es ist erkennbar, weshalb er dies tat, und konsequent. Aber trotz der Assoziation der Wörter, Paradies und Parodie, hat es natürlich Sinn, von einem Paradies der Mathematiker zu sprechen, weil sie dort am weitesten von der Sünde des bloßen Rechnens und seiner Fortsetzungen entfernt sind aber von der Parodie zu sprechen und dazu noch anzunehmen, sie sei als solche verkannt und nachträglich dem ordentlichen Bestand ern−//ster mathematischer Verfahren einverleibt worden, setzt den Willen zur drastischen Außenseitigkeit voraus. Ein Satiriker verhöhnt die Mathematiker und wird daraufhin von diesen zum Mathematiker ernannt. Das erinnert an Verfahren, die sonst nur in der modernen Ästhetik möglich sind: ein Produktlspricht jeder Nähe zur Kunst Hohn, und gerade deshalb findet es sich eines Tages als Kunstwerk im Museum wider. Der gemeinsame Kunstgriff ist, die Herkunft und Absicht unsichtbar zu machen." The symbol // denotes that the text continues on the verso side.

[23] This paragraph is not underlined in Blumenberg's exemplar of Wittgenstein (1969), but a small letter 'k' is written in pencil to the side of this paragraph.

paragraph before, set theory to the "invention of new games" (1969, p. 381), Wittgenstein calls for more philosophical clarity, which does not concentrate on the "infantile" (i.e. set theory), or on what is played at the dark. That is, Wittgenstein develops here a philosophy of mathematics that does not get entangled in the debate about the foundations of mathematics. It does not try to seek or justify the foundation of mathematics as the logicists, formalists, and intuitionists do, being the three major philosophical movements at the beginning of the twentieth century, all dealing, in one way or another, with the foundations of mathematics.[24] Instead, Wittgenstein argues that mathematics needs no foundation, but rather it is rooted in practice and the plurality of games.

Blumenberg stresses this citation, but not so much in order to enter a debate on the history of philosophy of mathematics, but rather to claim a somewhat opposite, or at least more nuanced, claim about mathematical practices: Blumenberg agrees that philosophy's primary effect is to "make the sprouts wither" (1989, p. 790), but he notes that philosophy "tames the energy of great expectations, the disappointment of which can lead to world anger" ["[die Philosophie] bändigt die Energie der großen Erwartungen, deren Enttäuschung zum Umschlagen in Weltzorn verführen kann"] (1989, p. 791). This 'taming' of the energy can be read as an implicit reference to Freud, whom both Wittgenstein and Blumenberg mention with respect to the play with set theory, which is presented as a return to the mathematician's childhood. In this context, another card from Blumenberg's *Zettelkasten* 22, titled "LOG WITTGENSTEIN: NACH DER METHODE FREUDS: ZURÜCK AUFS INFANTILE,"[25] states sarcastically the mathematician's position on the foundations of mathematics: "In the language of Freud, the mathematician suffers from a complex that disgusts him with everything that concerns the early history of his own ideas, their childhood, their as yet unconsolidated, i.e. not yet exclusively object-oriented state" ["Der Mathematiker leidet, in der Sprache Freuds, an einem Komplex, der ihm alles verekelt, was die Frühgeschichte seiner eigenen Vorstellungen betrifft, ihre Kindheit, ihren noch nicht gefestigten, also noch nicht ausschließlich objektbezogenen Zustand"]. In this sense, the philosophy of mathematical practices can illuminate the dark places and the underground paths of mathematics itself. Calling for a plurality of approaches, the infantile would lose its meaning as such, and the 'complex' of the mathematician would be dissolved. Such philosophy enables to shed light—to use another metaphor of light and darkness—on other spaces of mathematical practices, which are beyond the logical one.

---

[24] See also Pérez-Escobar: "philosophy of mathematics has been characterized by an obvious and almost exclusive attention to the foundations of mathematics […]. Foundational studies often deal […] with questions on apriorism and logic or framed as ontological inquiries […]. [Hence] it [philosophy of mathematics] became a place for testing large classical philosophical views in metaphysics and epistemology" (2022, pp. 158–159).

[25] Blumenberg, Hans: LOG WITTGENSTEIN: NACH DER METHODE FREUDS: ZURÜCK AUFS INFANTILE, DLA Marbach, Bestand A: Blumenberg, Hans, Zettelkasten 22. The card is not numbered. On the verso side, one finds the following title, in the left upper corner "1/82 HGL VI 9."

5 Spaces of Mathematical Practice III: Blumenberg on "Der Mathematiker erzeugt *Wesen*"

The previous section examined Blumenberg's reflections in "Im Fliegenhaus." However, one may examine the mathematical puzzle (of the reconstruction of the rectangle), presented by Wittgenstein in BGM and reconceptualized by Blumenberg, in another way, concentrating on Blumenberg's cards. To recall, Blumenberg analyzes this example during his reading of BGM; in his exemplar, the relevant sections to this example are found on pages 55–58. Notably, several pages earlier, on page 50, Blumenberg underlines another sentence: "The mathematician creates *essence*" ["Der Mathematiker erzeugt *Wesen*"] (Wittgenstein, 1974, p. 50). Next to this underlined sentence, the letter 'k' is written, indicating that Blumenberg wrote a card associated to it.

Before analyzing this card, let us examine this section in Wittgenstein's BGM:

> I might also say as a result of the proof: 'From now on an H and a P are called 'the same in number'. Or: The proof doesn't *explore* the essence of the two figures, but it does express what I am going to count as belonging to the essence of the figures from now on. – I deposit what belongs to the essence among the paradigms of language. The mathematician creates *essence* [*Wesen*] (1974, p. 50).

Wittgenstein proposes to consider here the proof, seen as a linguistic means, as what expresses a future essence. In this view, the essence of the mathematical object is not predetermined (because "the proof doesn't *explore* the essence of the two figures"), but it rather can be always changed and re-determined. This perspective underpins Wittgenstein's statement that the "mathematician creates *essence*."

How did Blumenberg read this passage? In September 1984 Blumenberg wrote the following card concerning this section, connecting it with a former section from BGM. The card, titled "POS WTG: "URSTIFTUNG": MATHEMATIK SCHAFFT WESEN = SEKUNDÄR PLATON,"[26] is as follows:

> Wittgenstein, WW VI 50:
>
> "The mathematician creates <u>essence</u>."
>
> Counter-thesis to the thesis of Russell: He says that the final proposition in an inference always already follows [from the previous propositions], and it only needs to be inferred, as Frege had said, the straight line connecting two points is actually already there before it is drawn. (45) Note that we are not dealing here with ideality in the classical sense, for so that a proposition may follow from others even before it is inferred, these factual and arbitrary propositions, which are not linked with any ideality, must be there. It is the same with the straight line between two points: it is already there before it is established, but it cannot be there before those two points are factually there whose position has nothing of ideality, so that [one] could speak of a secondary ideality that results from primary facts and only means the relationship that always already exists between these facts, regardless of whether

[26] Blumenberg, Hans: POS WTG: "URSTIFTUNG": MATHEMATIK SCHAFFT WESEN = SEKUNDÄR PLATON // PHA IV (num. 023977), DLA Marbach, Bestand A: Blumenberg, Hans, Zettelkasten 11: P–S. On the verso side it is written in red ink: "9/84."

it is established. // To this, Wittgenstein's laconic proposition of the generation of the essence, i.e. of a primordial foundation, is the harshest antithesis.

Wittgenstein compares Frege's statement that a construction line exists in a certain way before it is drawn with the pre-existence of numbers, including those that are counted, written, and thought. It is the 'primal experience' of Platonism written down: "How strange this feeling, as if they were all already somehow present!" (WW IV 281).[27]

What does Blumenberg claim here? Let us first note that with '(45)' Blumenberg refers to page 45 in Wittgenstein's BGM, and specifically to the following section there, when the underlined sentences are by Blumenberg:

> In his fundamental law, Russell seems to be saying of a proposition: 'It already follows—all I still have to do is, to infer it'. Thus, Frege somewhere says that the straight line that connects any two points is really already there before we draw it; and it is the same when we say that the transitions, say in the series + 2, have really already been made before we make them orally or in writing—as it were tracing them (Wittgenstein, 1974, p. 45).[28]

Such reflections are not to be found in the *Höhlenausgänge*. Nevertheless, they indicate not only Blumenberg's agreement with Wittgenstein but also his own reflections concerning the philosophy of mathematics and the nature of mathematical objects. Blumenberg in fact presents such reflections in the last pages of his book *Lebenszeit und Weltzeit*, but the above card offers a more elaborate discussion. However, in *Lebenszeit und Weltzeit* Blumenberg stresses the "dynamic Platonism" found in Wittgenstein's thought, a term not found in the above card (1986, p. 374).

---

[27] "'Der Mathematiker erzeugt Wesen.' Gegenthese zu der These von Russell, der von dem Schlußsatz in einer Folgerung sagt, er folge immer schon, brauche nur noch gefolgert zu werden, wie Frege gesagt hatte, die Gerade, die zwei Punkte verbindet, sei eigentlich schon da, ehe sie gezogen würde. (45) Man beachte, daß es hier nicht um Idealität im klassischen Sinne geht, denn damit ein Satz aus anderen folgen kann, noch ehe er gefolgert sei, müssen diese faktischen und beliebigen, mit keiner Idealität zu verbindenden Sätze da sein. Ebenso ist es mit der Geraden zwischen zwei Punkten: sie ist schon da, bevor sie hergestellt wird, aber sie kann nicht da sein, bevor nicht jene beiden Punkte faktisch da sind, deren Lage nichts von Idealität hat, so daß [man] von einer sekundären Idealität sprechen könnte, die sich aus primären Fakten ergibt und nur das Verhältnis meint, das zwischen diesen Fakten immer schon besteht, unabhängig davon, ob es festgestellt wird. // Dazu nun ist Wittgensteins lakonischer Satz von der Erzeugung des Wesens, also von einer Urstiftung, die härteste Antithese.

Wittgenstein vergl Freges Ausspruch, eine Konstruktionslinie sei in gew Weise vorhanden, bevor sie gezogen werde, mit dem Vorvorhandensein [Vorhandensein] der Zahlen, auch solcher, die die gezählt, geschrieben, gedacht werden. Es ist das 'Urerlebnis" des fortgeschriebenen Platonismus: ‚Wie seltsam dieses Gefühl, als wären sie schon alle irgendwie vorhanden!'(WW IV 281)."

In the first paragraph, Blumenberg adds the word 'man' at the bottom of his card. The reference at the end of the card is to Wittgenstein (1969, p. 281). In his exemplar, Blumenberg underlines the cited sentences. The last paragraph of the text on the card is indented and typed in another font.

[28] "Russell scheint mit jenem Grundgesetz von einem Satz zu sagen: ‚Er folgt schon—ich brauche ihn nur noch zu folgern.' So heißt es einmal bei Frege, die Gerade, welche je zwei Punkte verbindet, sei eigentlich schon da, ehe wir sie zögen und so ist es auch, wenn wir sagen, die Übergänge, der Reihe +2 etwa, wären eigentlich bereits gemacht, ehe wir sie mündlich oder schriftlich machen, – gleichsam nachzögen."

While one may argue that Blumenberg's emphasis is very similar to Wittgenstein's, there are two key differences. The first concerns the emphasis that Blumenberg puts on the 'secondary Platonism,' an expression that does not appear in BGM. Here 'secondary' does not mean the opposite of 'primary,' as something less important, as if the secondary topic is subjected to a more primary one, whose traits are already known. Rather, it denotes the reverse: exactly because of the discovery of new properties of mathematical objects or previously unknown relations between such known objects, the essence of these objects changes retroactively. When Blumenberg calls Wittgenstein's claim a Platonist one, he radically reinterprets Platonism: the traditional platonic character of the mathematical objects, which is assumed by Russell and Frege, is replaced by a 'Platonism' that allows a possible and constant change in the meaning of mathematical objects and concepts. In this sense, the antithesis to Russell is indicated when Blumenberg mocks Frege's claim that the objects were, strangely enough, always there as 'somewhat given.'

The second difference lies in how Blumenberg notes the 'ideality' of the mathematical objects, and how this 'ideality' is perpetually undermined. It is challenged by new proofs (and solved problems), but such proofs—as indicated by Wittgenstein—can also be regarded as linguistic operations. Pointing out that the ideality of the mathematical object in fact does not exist, Blumenberg stresses, even if implicitly,[29] that mathematical concepts are always subject to change *because* they are subjected to and created by language. In this context, the "ideal of mathematics", which is bound to "the desire for linguistic definiteness"—so Blumenberg in 1975 in his notes on non-conceptuality (2020, p. 282)—can never be reached.

6 Conclusion: A Call to Mathematical 'Controlled Ambiguity'?

How can one extract from Blumenberg's reading of Wittgenstein's writings a coherent position concerning mathematical objects and concepts, and, more generally, concerning various views in the philosophy of mathematics? Here one may follow Elise Marrou (2013), who emphasizes the notion of 'plurivocité contrôlée' ['controlled ambiguity'] in Blumenberg's thought already in 1966. Blumenberg stresses that the

> philosophical language [...] I would describe its tendency as directed towards 'controlled ambiguity' [*kontrollierte Mehrdeutigkeit*]. [...] This tendency is based on the utilization of mundane language for the designation of transcendental concepts, for there obviously can-

[29] Implicitly, because Blumenberg does not cite Wittgenstein's claim, that "essence [is deposited] among the paradigms of language" (1974, p. 50).

not be a specifically transcendental language, just as the illusion of a philosophical language of its own has been fading since the late Wittgenstein (2001, p. 125).[30]

In this sense of controlled multiple options of making sense, one can also interpret Blumenberg's call to consider 'secondary Platonism' as a philosophical position regarding mathematics.[31] Such secondary Platonism may imply an opposite conception of Wittgenstein, to be found in his last work *Über Gewißheit*. There Wittgenstein calls mathematical theorems fossils: "Petrefakten" (1970, p. 169).[32] However, Blumenberg's call to see mathematical objects as embedded in such secondary Platonism prompts to view such mathematical theorems as what may act on and change retroactively mathematical concepts; such fossils are therefore not dead objects. Thus, the conception of secondary Platonism and the emphasis on '*kontrollierte Mehrdeutigkeit*' are linked not only to the tendency of 'philosophical language' but also to mathematical language.

The concept of 'controlled ambiguity' also stresses the open-endedness of mathematical objects. As both Wittgenstein and Blumenberg note, even the solution to simple mathematical puzzles can prompt a renewed consideration of the concept of 'rectangle,' leading to an open-ended process of redefining again and again mathematical objects (Pérez-Escobar, in press). Such ambiguity in mathematics, which is on the one hand unexpected and on the other hand forces itself, can be described, following Marrou, as "post-cartesian rationality" (2013, p. 9; 2016, p. 115). This type of rationality also reflects the role of the mathematician who "creates essence" ("erzeugt Wesen") (Wittgenstein, 1974, p. 50), as the creator of new possibilities, as the one who calls for a plurality of options, hence embracing various practices and not necessarily focusing (only) on the foundations of mathematics. This is the reason why set theory (and, in general, the philosophical debate on mathematical foundations) might be viewed as a parody. Not (only) because one returns to 'childish games,' but (also) because one recognizes the discourse on foundations and the

---

[30] "philosophische Sprache [...] [i]hre Tendenz möchte ich als auf‚kontrollierte Mehrdeutigkeit'gerichtet bezeichnen. [...] gründet sich diese Tendenz auf die Indienstnahme der mundanen Sprache für die Bezeichnung transzendentaler Begriffe, denn eine spezifisch transzendentale Sprache kann es offenkundig nicht geben, wie überhaupt die Illusion einer philosophischen Eigensprache seit dem späten Wittgenstein zerrinnt."

[31] Though outside the scope of this chapter, one should note that a similar view concerning 'productive ambiguity' in mathematics is to be found in Grosholz (2007).

[32] See Wittgenstein (1970, p. 169): "Die Sätze der Mathematik, könnte man sagen, sind Petrefakten. —Der Satz 'Ich heiße ...' ist dies nicht. Aber von denen, die, wie ich, die überwältigende Evidenz haben, wird auch er als *unumstößlich* betrachtet. Und das nicht aus Gedankenlosigkeit. Denn, daß die Evidenz überwältigend ist, besteht eben darin, daß wir uns vor keiner entgegenstehenden Evidenz beugen *müssen*. Wir haben also hier einen Widerhalt ähnlich wie den, der die Sätze der Mathematik unumstößlich macht" ["The propositions of mathematics might be said to be fossilized. —The proposition 'I am called ...' is not. But it too is regarded as incontrovertible by those who, like myself, have overwhelming evidence for it. And this not out of thoughtlessness. For, the evidence's being overwhelming consists precisely in the fact that we do not need to give way before any contrary evidence. And so we have here a buttress similar to the one that makes the propositions of mathematics incontrovertible"].

purely logical space of formalization as ridiculous. Such recognition opens a door or another space to other mathematical practices, such as diagrammatical ones, that do not necessarily let themselves be formalized. By considering the ways in which Blumenberg reacted to Wittgenstein—via the published writings and the private cards—one sees that Blumenberg does not necessarily reject the optimism and trust in the paradise represented by Hilbert's approach. Instead, Blumenberg joins Wittgenstein to inspect closely mathematical caves. Blumenberg, in this sense, points towards the plurality of options incorporated in the mathematical '*kontrollierte Mehrdeutigkeit.*'

References

Blumenberg, H. (1981). Lebenswelt und Technisierung unter Aspekten der Phänomenologie. In *Wirklichkeiten, in denen wir leben* (pp. 7–54). Reclam.

Blumenberg, H. (1986). *Lebenszeit und Weltzeit*. Suhrkamp.

Blumenberg, H. (1987). *Die Sorge geht über den Fluss*. Suhrkamp.

Blumenberg, H. (1989). *Höhlenausgänge*. Suhrkamp.

Blumenberg, H. (1997a). *Die Vollzähligkeit der Sterne*. Suhrkamp.

Blumenberg, H. (1997b). *Shipwreck with spectator: Paradigm of a metaphor for existence*. MIT Press.

Blumenberg, H. (2000). *Die Verführbarkeit des Philosophen*. Suhrkamp.

Blumenberg, H. (2001). Sprachsituation und immanente Poetik. In *Ästhetische und metaphorologische Schriften* (pp. 120–135). Suhrkamp.

Blumenberg, H. (2007). *Theorie der Unbegrifflichkeit*. Suhrkamp.

Blumenberg, H. (2010). *Paradigms for a Metaphorology*. Cornell University Press.

Blumenberg, H. (2020). Theory of nonconceptuality. In *History, metaphors, fables. A Hans Blumenberg reader* (pp. 259–297). Cornell University Press.

Ferreirós, J. (2007). *Labyrinth of thought: A history of set theory and its role in modern mathematics*. Birkhäuser.

Ferreirós, J. (2008). The crisis in the foundations of mathematics. In T. Gowers (Ed.), *The princeton companion to mathematics* (pp. 142–156). Princeton University Press.

Friedman, M. (2024). On metaphors of mathematics: Between Blumenberg's nonconceptuality and Grothendieck's waves. *Synthese, 203*, 149. https://doi.org/10.1007/s11229-024-04558-3

Grosholz, E. R. (2007). *Representation and productive ambiguity in mathematics and the sciences*. Oxford University Press.

Haverkamp, A. (2018). *Metapher – Mythos – Halbzeug: Metaphorologie Nach Blumenberg*. Walter de Gruyter.

Hertfelder, K. (2022). Zur Formel bei Hans Blumenberg. In H. Bajohr & E. Geulen (Eds.), *Blumenbergs Verfahren* (pp. 233–267). Wallstein.

Hilbert, D. (1926). Über das Unendliche. *Mathematische Annalen, 95*(1), 161–190.

Kambartel, F. (1963). Philosophische Perspektiven der Diskussion um die Grundlagen der Mathematik. *Archiv für Geschichte der Philosophie, 45*, 157–193.

Kambartel, F. (1968). *Erfahrung und Struktur. Bausteine zu einer Kritik des Empirismus und Formalismus*. Suhrkamp.

Kroß, M. (2010). Metapher als Entzug—Metapher als Gabe: Wittgenstein mit Blumenberg lesen? In V. Munz (Ed.), *Essays on the philosophy of Wittgenstein* (pp. 75–92). Walter de Gruyter.

Marrou, É. (2013). Aspects de la plurivocité contrôlée: Portrait de Wittgenstein par Blumenberg. *Implications philosophiques*. https://www.implications-philosophiques.org/aspects-de-la-plurivocite-controlee-portrait-de-wittgenstein-par-blumenberg/

Marrou, É. (2016). Autoportrait de Blumenberg en thérapeute viennois. De nobis ipsis silemus. *Archives de Philosophie, 79*(1), 109–120.
Menke, B. (2009). Sumpf und Mauer. Versuche zu einer Philosophie der Unbestimmtheit. In A. Haverkamp & D. Mende (Eds.), *Metaphorologie Zur Praxis von Theorie* (pp. 316–338).
Nicholls, A. (2014). The Goethe complex: Hans Blumenberg on Das Dämonische. In E. Geulen, K. Wetters, C. Meierhofer, & L. Friedrich (Eds.), *Das Dämonische. Schicksale einer Kategorie der Zweideutigkeit nach Goethe* (pp. 97–119). Fink.
Nietzsche, F. (1923). *Gesammelte Werke, Achter Band. Menschliches, Allzumenschliches.* Musarion.
Pérez-Escobar, J. A. (2022). Showing mathematical flies the way out of foundational bottles: The later Wittgenstein as a forerunner of Lakatos and the philosophy of mathematical practice. *KRITERION – Journal of Philosophy, 36*(2), 157–178. https://doi.org/10.1515/krt-2021-0041
Pérez-Escobar, J. A. (in press). The open-endedness of mathematical puzzles and its enculturated control: Navigating necessary tensions. In D. Sarikaya (Ed.), *From mathematical riddles to research: What makes a problem good?* Birkhäuser.
Wittgenstein, L., (1963). *Schriften – Tractatus logico-philosophicus. Tagebücher 1914–1916. Philosophische Untersuchungen.* Suhrkamp.
Wittgenstein, L. (1969). *Schriften 4. Philosophische Grammatik.* Suhrkamp.
Wittgenstein, L. (1970). *Über Gewißheit.* Suhrkamp.
Wittgenstein, L. (1974). *Schriften 6. Bemerkungen über die Grundlagen der Mathematik.*.
Wittgenstein, L. (1978). *Schriften 7. Wittgensteins Vorlesungen über die Grundlagen der Mathematik, 1939.* Suhrkamp.
Zill, R. (2008). 'Sagen, was sich eigentlich nicht sagen läßt'. Theodor W. Adorno, Hans Blumenberg und andere Leser Wittgensteins. In E. Alloa, & A. Lagaay (Eds.), *Nicht(s) sagen. Strategien der Sprachabwendung im 20. Jahrhundert* (pp. 41–60). transcript.
Zill, R. (2020). *Der absolute Leser: Hans Blumenberg. Eine intellektuelle Biographie.* Suhrkamp.

Michael Friedman is a research associate at the Hausdorff School for Mathematics at the University of Bonn and a Senior Lecturer at the Cohn Institute For History And Philosophy of Science And Ideas at Tel Aviv University. The focus of his research is on how material, visual and symbolical knowledge interact with each other. In this context, he looks into the material practices of mathematics (folding, weaving, braiding, knotting, as well as 3D models) and how symbolical-mathematical knowledge was prompted by them. He also researches Hans Blumenberg's reflections on the various metaphors and concepts of mathematics. Latest publications: *On Joachim Jungius' Texturæ Contemplatio. Texture, Weaving and Natural Philosophy in the 17th Century* (2023); *Ramified Surfaces. On Branch Curves and Algebraic Geometry in the 20th Century* (2022).

Part II
Phenomenology as Anthropology

Crisis and Restitution. Phenomenological Subjectivity as Root of Blumenberg's Description of Historical Change

Pedro García-Durán

Abstract This chapter explores the philosophical ideas and descriptions that Blumenberg derived from the phenomenological descriptions of intentional consciousness for his understanding of history. It focuses on the significance of Husserl's contributions which form the basis of books such as *The Legitimacy of the Modern Ages* and *Work on Myth*, demonstrating how they provide a structural foundation that renders historical descriptions coherent and imboues them with a deep philosophical relevance. Therefore, this writing aims to investigate the influence of Husserl rather than Blumenberg's adherence to the phenomenological method. To achieve this objective, the chapter commences by analyzing the conclusions of his posthumous phenomenological anthropology and its compatiblity with historicity. This section will also elucidate the significance of the anthropological grounding of the intentional subject as exposed in *Beschreibung des Menschen*. Subsequently, the chapter delves into how the idea of historical crisis as it is studied in *The Legitimacy of the Modern Ages* aligns with these descriptions of intentionality and its mechanisms of restitution. Lastly, the chapter explains the way in which the philosopher from Lübeck describes the legitimacy of history as a consequence of these procedures and characteristics of subjectivity identifying historicity with humanity. In conclusion, this chapter tries to unveil some of the phenomenological influences that endow Blumenberg's thought with exceptional philosophical value.

This work was made with the help of the research project *Historia conceptual y crítica de la modernidad* (FFI2017-82195-P), funded by MCIN/AEI/10.13039/501100011033 and by FEDER, A Way of Making Europe, and the eponymous research group at the University of Valencia (GIUV2013-037).

P. García-Durán (✉)
Independent Scholar, Universitat Jaume I of Castelló, Castelló, Spain
e-mail: pm.garciaduranjime@edu.gva.es

A. Fragio et al. (eds.), *Hans Blumenberg's History and Philosophy of Science*, Studies in History and Philosophy of Science 63,
https://doi.org/10.1007/978-3-031-78251-0_4

1 Introduction

This presentation delves into the phenomenological roots of Hans Blumenberg's historical narratives, particularly exploring the parallels between his views on history and change and the descriptions of intentional consciousness in phenomenology. This analysis sheds light on the pervasiveinfluence of phenomenology in Blumenberg's work, an influence that is not limited to the books and papers in which he discusses directly with members of that intellectual school, but underlays the totality of his production.[1] It goes beyond a mere philological observation to uncover the epistemological foundations of his historical work. Furthermore, as Blumenberg seldom discusses these assumptions in his historical writings, its exposure contributes to understand his philosophical aim and scope, and, therefore, to discuss its present validity and its possible actualizations.

To do this, I will begin by clarifying the implications of these phenomenological borrowings in Blumenberg's thought, showing the magnitude of phenomenology in his work, which goes beyond these influences and becomes a real way of living and thinking.[2] These explanations will set the context from which the aforementioned descriptions are taken and will allow us to explain how the understanding of intentionality is shaped in an outline that fits with his description of the historical run of mankind. After making this intentional 'skeleton' explicit, the chapter will explain its impact on his historical work, exposing how it defined the perspective of *The Legitimacy of the Modern Ages* and *Work on Myth*. These two books were distant in time (their first editions were published in 1966 and 1979 respectively) and, for that reason, they are useful not only to show the permanence of the phenomenological influence in different periods of Hans Blumenberg's production, but also to illustrate the continuity that exists in his thought. Finally, to close my contribution, I will discuss how this intentional scheme can still be fruitful as a historical descriptive method.

[1] Blumenberg's phenomenology is the subject of my book *El camino filosófico de Hans Blumenberg* (García-Durán, 2017; see also 2015). However, this chapter is not focused on the phenomenological praxis that this philosopher applied to his research as the book was, but on the reception of Husserl's descriptions of subjectivity and its features. Even if the relations between the German philosopher and the phenomenological school are a main issue in his work, the bibliography on the subject is still scarce. There are, nonetheless, some interesting references such as Hernández Marcos (2015), Lederle (2021), Merker (2009), Philippi (2019), Stoellger (2000), Zowislo (2009), Zambon (2019), or the number 2023/2 of *Phänomenologische Forschungen* (Buch & Zambon, 2023).

[2] As Blumenberg wrote to an American colleague about his philosophical influences: "My heart and my intellect belong to Kant, but what rest has he left me when I have understood him? An exegetical life. His inheritance is certainly not a method, because "his" philosophy has not provided it: the *Critiques* are, systematically considered, preparatory, and do not allow us to glimpse what the 'doctrine of the method' would look like […]. Can you live with Kant? No. But you can live with Husserl" (quoted in Fragio Gistau, 2012, p. 664).

2 Blumenberg and Phenomenology

As I have pointed out, Blumenberg's relationship with phenomenology is a constant in his work. However, this interconnection is not unambiguous, nor does it lack changes over time. Some interpreters have highlighted the circularity of the explicit discussions with phenomenological topics in Blumenberg's intellectual biography, as his main phenomenological works are dated in his early years (especially his *Habilitationschrift*) and his later years.[3] Nevertheless, the approach to this topic may seem opposite between those periods. While *Die ontologische Distanz* issues what may look like a death certificate for the phenomenological school and the philosophical tradition from which it came, the later works not only discuss with Husserl's thought as a valid interlocutor, but also try to provide a plausible anthropological foundation for its descriptions of consciousness.[4] As paradoxical as this transformation seems, it may have a simple explanation. When Blumenberg did his scholarly research in 1950, Husserl's huge archive was almost undiscovered. The publication of *The Crisis of the European Sciences and the Transcendental Phenomenology* in 1954[5] meant a revolution for the reception of the Moravian thinker, as it described thoroughly the functions of the lifeworld, probably the most successful concept in Husserl's late work.[6] This terminological finding soon became a cornerstone of Blumenberg's thought[7] and allowed him to create, only 5 years later, an approach to history that he labeled as a 'phenomenology of history.' A

[3] Before the author's death, this return to phenomenology was already explicit in *Lebenszeit und Weltzeit*. This influence is much more visible nowadays, especially thanks to the research on his *Nachlass* and a relevant number of posthumous books such as *Zu den Sachen und zurück* (2002), *Beschreibung des Menschen* (2006), *Theorie der Lebenswelt* (2010), *Phänomenologische Schriften* (1981–1988) (2018) or *Die ontologische Distanz. Eine Untersuchung zur Krisis der philosophischen Grundlagen der Neuzeit.* Habilitationsschrift of 1950 (2022). There are two early and remarkably interesting writings on this topic, both published in the Italian monograph *Hans Blumenberg. Mito, metafora e modernità* (Accarino, 1999; Carchia, 1999). However, the usual interpretation of Blumenberg's relation with phenomenology highlights the writings on which the philosopher from Lübeck addresses it directly and overlooks the permanence of a methodological and descriptive legacy throughout his entire work.

[4] As Blumenberg states in *Beschreibung des Menschen*: "A phenomenological anthropology would be the part of phenomenology that brings into evidence the phenomenologist's assumptions as anthropological assumptions or that at least provides them with a hypothesis that is difficult to improve" (2006, p. 166). Hans Blumenberg's quotes come from the English editions when they are available, otherwise, they will be my own translations from the German originals.

[5] The book appeared as the sixth volume of the *Husserliana* collection in 1954. There was a partial edition in 1936 in Belgrade which did not include the parts corresponding to the lifeworld.

[6] There are numerous books and papers on the concept and its subsequent influence in philosophy, human and social sciences. One of the most complete and interesting among them, even if it was published long ago, is *Der Begriff der Lebenswelt. Theorien vortheoretischer Erfahrungswelt* (see Welter, 1986).

[7] The first work by Blumenberg focused on the subject was *Lebenswelt und Technisierung unter Aspekten der Phänomenologie*, a conference held in 1959 that was published in the Italian magazine *Filosofia* four years later (see Blumenberg, 1981).

perspective that tries to "grasp the moving relation between the lifeworld and the concept of reality" (Jauss, 1969, p. 226).[8]

The Hanseatic thinker may have seen a main descriptive tool in this famous concept, but the importance of the Husserlian influence is not limited to such a theoretical borrow. What *Crisis* exposed was Husserl's concern about the dissolution of his once solid phenomenological circles. The renunciation to factuality implied by the transcendental reduction was the main reason that led to the dissolution of the school. The kind of mundane asceticism required by the transcendental turn of phenomenology was anachronistic after the so called 'masters of suspicion,' Evolutionism and the expansion of historical thinking. However, this approach was countered in *Crisis* with the genetic perspective that tries to give account of the lifeworld foundations of the pure essences inside consciousness.[9]

By doing that, Husserl made clear his commitment to the original phenomenological motto of going 'back to the things themselves' in the most accurate possible fashion. Although *Crisis* kept the restrictions of the transcendental methodology of phenomenology, it was an attenuation of the autonomy of reason, as it was always linked to the primary experience of the lifeworld. The success of this new concept was also another proof of the fruitfulness of the methodology and unveiled its true nature as a "flexible descriptive method" (Blumenberg, 2018, p. 13), rather than a transcendental speculation isolated from daily life as was considered when the *Ideas Pertaining to a Pure Phenomenology and to a Phenomenological Philosophy* first appeared in 1913. This may explain the importance of Blumenberg (2002) adscription to phenomenology as a way of work and, as we said, even, of life, which is clearly exposed in later works. This adherence is present in his books and papers of the '60s and '70s, because, even if neither the school, nor its contributions, are openly discussed in them, these publications can be understood as practical applications of a phenomenology of history as a methodological tool.

Thus, the work of Blumenberg can be read as a heterodox approach to phenomenology. However, this will not be the path we travel in this writing. Here, we will focus on the theoretical legacy, the borrowing of the results of this method, which are surely the main reason for the adscription to the school, but remain to some extent implicit in most of Blumenberg's historical books. However, this application of phenomenology will raise questions for anyone who is aware of Husserl's thought; especially regarding the reason why these concepts linked to the notion of intentionality are extrapolated by Blumenberg to concrete historical and anthropological subjects, which were explicitly prohibited from a strict phenomenological point of view.[10] The main reason for this was already made clear in the last pages of

[8]As Blumenberg expressed in his reply to Dieter Henrich objections in the debates of the first meeting of the group *Poetik und Hermeneutik*.

[9]This topic was well documented by Ludwig Landgrebe, a former Husserl's assistant and later Blumenberg's professor at the University of Kiel, whose influence on the German philosopher's early work is acknowledged. See the texts included in Landgrebe (1963).

[10]In the pages of *Crisis*, Husserl (1954) separates his point of view from the historian's perspective, pointing out that the latter must focus on the everchanging particular lifeworlds and not on their

Die ontologische Distanz, where he explains the scope of the death certificate that was issued there:

> The phenomenological analyses of the structures intricately linked to their concept of horizon do not lose their value or become false, but they lose their radical and universal claim, as far as they would have been highlighted in relation to a certain "place" of the ontological movement (Blumenberg, 2022, p. 266).

Therefore, the main cause of Blumenberg's distance from Husserl is the transcendental assumptions that sustain his system and not the philosophical results he accomplished. An ideal that, in the German philosopher's opinion, reached its end with the rise of historicism and anthropology in the second half of the nineteenth century. An event that led to the factualization of consciousness and its structures, since then inescapably linked to their socio-historical context (Blumenberg, 2022). However, the application of the descriptions to the historical constitution and changes may also show the validity of these contributions of the phenomenological school, at least as long as we accept Blumenberg's historical understanding. So, the main intention of this presentation is to show how history works as the result of the interaction between the intentional subject and its world. Nevertheless, it may seem paradoxical that such successful descriptions rely upon false foundations. This problem is also addressed in Blumenberg's work. As we have advanced, *Beschreibung des Menschen* gives an anthropological justification of intentional consciousness and its features. For that reason, this presentation varies the chronological order of Blumenberg's production. In the next part, I will try to outline his understanding of intentionality through his later works, especially the posthumous phenomenological anthropology where this problem is discussed in detail. An understanding that has already been used in earlier works. The phenomenological skeleton can be recognized in the historical examples of the passage from the Middle Ages to Modernity and in the emergence of the anthropological mechanisms developed to produce signification, which enabled humanity to create Mythology as is described in *Work on Myth*. This will clarify what kind of unity underlies Blumenberg's thought throughout his different periods and what philosophical and theoretical reasons lead to his attempt to understand the nature and conditions of possibility of historical change and its relationship with the construction of consciousness through a phenomenology of history.

---

value for the operations of consciousness as a whole. On the other hand, the subject of the phenomenological prohibition of anthropology is deeply studied in *Beschreibung des Menschen* (Blumenberg, 2006).

3 Anthropological Intentionality

For Blumenberg, the most defining anthropological feature was visibility. Bipedalism and the change from the jungle to the open horizons of the savannah turned vision into the most important among the human senses; not only because this new position opened the visual range, but also because hominids became more visible to possible threats too. This apparently simple idea has many consequences for the German philosopher that explain and complement the concept of intentionality. As Blumenberg confirms (2006), all the metaphors and descriptions of intentional consciousness are related to a visual approach to the relation between *cogito* and *cogitata*. The metaphor of the horizon can be the main example of this dependence because it implies that phenomena do not appear isolated but are placed in the middle of a network of relations, which must be organized and determined to be able to isolate the particular content of consciousness that needs to be analyzed. The first steps of mankind, as Blumenberg describes them, may have consisted of the construction of a reliable intellectual order to contain a still formless experience, the creation of a safe place from which he could focus his attention on the outside. It was the process of building what Blumenberg would call 'cultural cases' where humanity could develop its living.[11]

However, experience is fluid. The net of relations in which the phenomena are embedded is not only deployed in space, but it also flows in time. In fact, for phenomenology, consciousness is equivalent to time and, for that reason, the Husserlian descriptions of inner consciousness of time are a centrepiece of Blumenberg's interest in phenomenological philosophy.[12] However, these descriptions expose the weaknesses of the transcendental methodology as they rely on the ability of reflectiveness and, thus, not on the pure presence of the phenomena. Inner experience of time can only be addressed as a memory of the temporal integration of realities, and a memory is not a phenomenon; it is not comparable to pure presence. That is why Blumenberg accuses Husserl of deducing and not describing his famous understanding of temporality.[13] Memories cannot be verified without access to factuality, as their reliability is assessed by the witnesses that shared them and by their effects on our daily experience. Therefore, their consistency cannot be analyzed from the transcendental point of view. But, even if his assumptions mislead Husserl, he also manages to give a valid account of the integration of present, past, and future in the

[11] Blumenberg develops this scenario in the anthropogenetic narrations of *Work on Myth* (1985b) and *Höhlenausgänge* (1989).

[12] Blumenberg addresses this issue in one of his later works, *Lebenszeit und Weltzeit*, but it is also a topic in *Beschreibung des Menschen* (2006) and in many other short writings published in the aforementioned posthumous books.

[13] "The phenomenological analysis of the intuition of time is under suspicion of being a description only on the surface, and a deduction in fact: an expansion of what resides and is present in the thing. This statement is so serious for the solidity of phenomenology that one would like to wish it to be doubted" (Blumenberg, 2006, p. 42).

phenomena, which points out the instability of experience even if he was not able to fully understand it or how its integration procedures work.

Succession is the key to temporality and, if time is the main feature of consciousness, there must be a unifying force which may be able to overcome the threats of a dissolution of the horizon of experience, such as the disruptions in the continuity of the flowing experience of the subject. For Blumenberg, this ability of consciousness is crucial to ensuring its persistence in the flow of the changing horizons towards which intentionality is directed. Describing "consciousness as self-repairing machine" (Blumenberg, 2018, p. 15) means that it can autonomously restore most interruptions in the everchanging flow of phenomena and horizons that may threaten its self-preservation as a continuum. The importance of the temporality of consciousness for understanding historicity cannot be overlooked, not only as a necessary metaphor for every explanation of a succession, but also as the place where the problem of history finds its template in memory:

> The problem of memory is that of history *in nuce*: How is it possible to sustain intentionality beyond the discrete states of a subject, of many subjects, of many generations and times? […] Time enables the experience of non-identity in identity as a presupposition that an identity can be understood as prior to non-identity (Blumenberg, 2006, p. 136).

The problem of continuity of consciousness is explained in Husserl's work through "genetic logic" (Blumenberg, 2006, p. 71ff.).[14] His research on this matter exposes how consciousness developed mechanisms to overcome the negation of the fulfilment of its intentionality and to restore continuity. The projections of intentional consciousness may be deceived or contradicted by experience, leading to the initiation of processes that allow the integration of these obstacles into the uniformity of the subjective flow. This procedure explains the evolution of reason and logic, as these disruptions provide opportunities for learning the mechanisms to solve them. This may have been the origin of the logical operation of negation. From a phenomenological point of view, this is not a historical fact, as it occurs in pure consciousness and not in the real world, but it provides a valuable procedure to analyse the course of history. Blumenberg will take this genetic perspective to his descriptions of history because whenever an idea is created, a step which cannot be reversed is taken:

> The process of the formation of historical identity is a process of ideation: we can only continue what, beyond the factual testimonies, we can turn into an idea, that to which the past production pointed by its intention. Consciousness is as such a structure of possible identity (2006, p. 533).[15]

Therefore, it can be said that reason is the main tool of humanity destined to maintain the continuity of intentional consciousness when it is threatened. Nevertheless, intentionality requires a foundation from which to be directed, a solid

[14] Blumenberg's source for these interpretations is Husserl's *Erfahrung und Urteil*, a book first published in 1939.

[15] This is also the subject of "Die Urstiftung" the last part of *Lebenszeit und Weltzeit* (Blumenberg, 1986).

background on which to stand firmly and point its gaze towards the defined horizon. This can be the role of the lifeworld as an unquestioned background of the experience of consciousness. The historical use of this concept in the *Crisis* book makes it a sort of criterion and a model on which theory must be shaped to be meaningful (Husserl, 1954). The cause of the crisis of the European sciences that Husserl described was precisely the loss of the link between theory and the lifeworld produced by the excess of formalization of the natural sciences in the Modern Ages. Nevertheless, Husserl addresses this pre-theoretical substrate as a constant form which underlies all conscious life, it describes a primary experience that lies before and behind the course of history, but it remains as a stable, closed, and cohesive sphere in the experience of consciousness even if the elements that comprise it change.

This scheme may grasp the image of history that can be found in *Crisis*. However, Husserl's transcendental demands left him with an idealistic teleological model. For Husserl, history is a totality as a "hidden unit of an intentional interiority" (1954, p. 74). From his point of view, history is the deployment of theory as the fulfilment of the ideal of autonomous reason which was created already in a particular moment of history and caused by its factual succession of events and ideas.[16] The phenomenological reduction closes any access to exogenous motivations and, thus, it is unable to give an account of the necessities that forced mankind to begin making history and, of the instability of his lifeworld and his horizons. This can only be done by taking factual historical realities into account. A step that a strict transcendental phenomenology cannot take. That return to the historical facts and their reflection on the subject would define the phenomenologically heterodox path followed by Blumenberg in his historical works.

4 Crisis and Restitution on Human History

This point of view can be recognized behind the description of the epochal threshold displayed in the *Legitimacy of the Modern Ages*. This book made Blumenberg known in the German academia and gave him a renowned voice in the debate on secularization.[17] Even if its main argument tries to overcome the implicit accusations that this concept casts on the Modern Ages, it also answers Husserl's criticism to the excessive formalization and technification of modern science and its consequences, especially the breaking of the links that connected theoretical knowledge

[16] For example, the idea of "to freely give to oneself, to one's entire life, rules founded on pure reason, taken from philosophy" (Husserl, 1954, p. 5) was the ideal that modern science has forgotten. However, the main reason for this mistake is that, for Blumenberg (2010), there is no way to address the question about the reasons which led to the first creation of theory from the phenomenological perspective to the first exit from the cave.

[17] The most complete book on the secularization debate is Monod (2002), a book written by a great Blumenberg reader and a former phenomenologist.

to the lifeworld providing it with signification. To some extent, Blumenberg may agree with the Husserlian diagnosis, but, unlike the founder of phenomenology, he will not understand the events which led to this new time as a deviation, but as a legitimate response to the threats and challenges of the dissolving medieval mentality.

For Hans Blumenberg, an epoch is not merely a name given to a fraction of time accurately dated; it is "not only a question of words, but also a question of things" (1985a, p. 8), a set of shared understandings and certainties which are active in individual consciousness and have an impact on daily life. This substratum may recall the functions of the lifeworld, as it is a preconceptual floor on which every event takes place and from where it gets its historical meaning. Individual events, such as the posting of the Wittenberg theses, make history only if they stand out in this underlying net (Blumenberg, 1985a).[18] Therefore, there would be some kind of epochal conscience which determines the possible impact of an event on the minds of its contemporaries. In that sense, the event studied in *Crisis* or in Blumenberg's famous book on Modern Ages, such as the legitimacy of modern knowledge, must be understood according to the needs and motivations that come from this preconscious experience. Husserl's transcendental point of view could not reach this level of factuality, thus, it failed to understand why the allegedly harmful detour of modern science was undertaken and made the whole endeavor a treason to the real and transcendental motivations of theory.

This motivational impulse can only be understood by analyzing the experience which led to this new model of knowledge. In other words, through a description of the transition from a crumbling temporal horizon to a new one. This period of crisis is referred to as the epochal threshold, and it is recognized by its loss of order (*Ordnungschwund*) (Blumenberg, 1962). The reason for this collapse is the instability of the medieval lifeworld or, in other words, the end of the security that this primary experience should offer. This situation arises from the development of internal contradictions into aporias, as Blumenberg stated: "Eras exhaust themselves more in the transformation of their certainties and unquestionable axioms into riddles and inconsistencies than in their solution" (1985a, p. 464). The experience of this epochal threshold is the focus of the last part of *The Legitimacy of the Modern Ages.* It is undoubtedly the cornerstone of the book's argument and its most complex and challenging part.

To overcome the condemnation of the Modern Ages as a unilateral subtraction of theological contents implied in the accusation of secularization, this work attempts to show its causes and specificity, the features that made the age an original answer to unanswered questions urgently awaiting for a response. The theoretical access to

[18] The first major historical works that Hans Blumenberg wrote, including the one mentioned and *The Genesis of the Copernican world* (published in 1975), share this perspective. In the latter, the necessity of this context is shown in the indifference towards the heliocentric models of Aristarchus and Nicholas of Oresme caused by the historical contexts in which they were conceived. It was necessary that the possibility of a Copernicus was opened and that could only happen after the end of the cosmical and theological models of transcendence.

that dimension of the experience is undoubtedly difficult. To prove the existence of an epochal conscience, Blumenberg compares two authors who lived at the poles of this threshold: Nicholas of Cusa and Giordano Bruno.[19] The first of them experienced the dissolution of a horizon and the other glimpsed the appearance of another. This situation defined their respective theoretical attitudes and motivational backgrounds. Thus, Cusa, even though he was able to point out new directions, will belong to a world in its last throes, and his desire to maintain the vanishing mentality will lead him to feel "abandoned to the dismemberment of life and thought" (Blumenberg, 1985a, p. 477). On the other hand, Bruno, despite his tragic and cruel fate, can transcend the horizon of the dying era and reformulate the theological contradictions in the search for a "harmonious solution" (Blumenberg, 1985a, p. 477).

Therefore, the difference between those positions cannot be found in the concepts and problems they address, but in their attitude towards them. In a very summary way, this difference can be exemplified in their use and understanding of Christology. Even though Nicholas of Cusa is considered a precursor of modernity due to his humanism and his epistemological proposal of *Docta ignorantia*, whose idea of conjectural knowledge is analogous to the hypotheses of modern science, his conscience would be theocentric. His concern for Christology, according to Blumenberg, will be the foundation of his trust in human beings, a dignity that legitimizes his conjectural knowledge as belonging to a chosen being: "Man is great because God is great" (1985a, p. 530). On the contrary, Bruno will be burned at the stake for attacking one of the basic principles of Christianity: the incarnation of Christ. A resignation dramatically expressed in his heretical rejection of the crucifix in the *Campo dei Fiore* before his execution. An attack that could only have been directed from a new vision of reality that excluded the irruption of any transcendental principle in the natural order regulated by merely immanent laws (Blumenberg, 1985a). That thought will lead Bruno to a rationalist pantheism, a way of thinking whose paths would determine the philosophical emptying of the idea of God in modern philosophy.

Giordano Bruno was already involved in the restitution of the epochal horizon, in the construction of a new godless world where everything must be explained immanently and whose sole driving force is that of human beings. As is known, self-assertion is the name given by Blumenberg to this specific feature of the nascent mentality which tries to describe the primary experience of this new life world:

> An existential program, according to which man posits his existence in a historical situation and indicates to himself how he is going to deal with the reality surrounding him and what use will he make of the possibilities that are open to him (1985a, p. 138).

---

[19] The importance of this part of the book for Blumenberg is evident from his extensive research on the work of these two philosophers. Prior to *The Legitimacy of the Modern Ages*, he worked on two excellent critical editions of works by both authors with extensive theoretical introductions: Cusa's *Die Kunst der Vermutung. Auswahl aus den Schriften* (published in 1959) and, three years after his work on the modern ages, he edited Bruno's *Das Aschermittwochsmahl* (published in 1969).

But even through these changes, there are continuities, elements that are transferred from one horizon to the other such as those identified as secularizations. Nevertheless, for Blumenberg, that concept implied a hidden reproach as it understood those transfers as an "illegitimate robbery of theological property" (1985a, pp. 47–48). In contrast, the philosopher believes that the transfers are unavoidable, as every new era must grow upon "the humus of its predecessor" (Blumenberg, 2022, p. 114). A new horizon can only be built with the materials left at hand after the collapse of the previous one, which leads to his famous idea of reoccupations.

Those continuities or reoccupations are the only way for the historian to produce an intelligible narration; they are the Ariadne's thread that allows him to unravel the labyrinth. As Blumenberg remarks:

> These examinations could not become experience of history if they did not satisfy the transcendental principle according to which the alteration of appearances refers us to something persistent. [...] This is why one of the preconditions for attaining clarity in relating doctrines to one another, and thus in differentiating them, is the possibility of demonstrating, through dissection, an identical fundamental system of elementary assertions needs, notions of the world and the self, on both sides of the threshold (1985a, p. 469).

Therefore, the crisis is described as an insurmountable loss of confidence in the lifeworld caused by the internal tensions of its components. It is the dissolution of a lifeworld and then of the concept of reality that emerged from it. This unsettling situation leads to the attempt to create a new horizon of stability to focus the efforts of humanity, to give them meaning. In that sense, the epochal change can be understood as the restitution of a lost consonance, which was one of the main functions of reason in phenomenological consciousness. As mentioned before, Blumenberg wrote some passages on "Consciousness and Its Ability for Restitution" in his later years, that may provide a key to understanding the way the reoccupation is produced, and the continuity it implies. The remains of the shipwreck are construction materials for the new ship because the fate of the sailor is to keep sailing. But does it follow a fixed course? Does navigation end in some safe harbour? And, especially, why did he start sailing? These questions cannot be answered in a book whose scope is the specificity of a peculiar time in history. However, these are questions that someone looking to understand history must try to answer. The need for a horizon of confidence, which remains unjustified here, does not explain the unity of history, *a conditio sine qua non* of any quest in such a search. Even the persistence of reoccupied ancient concepts, metaphors, myths or questions, their use in new horizons of meaning, does not imply any continuous intention throughout the times, and, hence, they are not able to legitimate the persistence of theoretical work, which, in my opinion, is one of the goals that Blumenberg's historical descriptions share with Husserl's *Crisis*. In conclusion, the questions on the actor of this history, the human being, remain unanswered, for this reason the needs and opportunities that made history possible must be found elsewhere. The solution to these shortcomings requires a different approach and explains the changes that have been pointed out by the interpreters in the work of Hans Blumenberg during the '70 s, namely the introduction of an anthropological and anthropogenetical dimension in his thought.[20]

5 Absolutism of Reality and the Necessity of Theory

Therefore, *The Legitimacy of the Modern Ages* focuses on a particular epochal change. It explains the questions and concerns that led to the collapse of a horizon and the corresponding institution of a new order, but it cannot solve the problems of a phenomenology of history as a whole that we have pointed out. The question about the needs that led human being outside their first lifeworld into a succession of temporary and unstable horizons or 'cultural cases' cannot be answered from within history as it is a question about the beginning of history itself. Thus, Blumenberg will try to answer it from anthropology, creating anthropogenetic narrations that explain such a passage according to the biological structure of human beings. In the first chapters of *Höhlenausgänge*, Blumenberg (1989) tells the story of the passage from the jungle to the savannah of the new biped animal, a change of ecosystem that places him in front of an undefined and threatening visual horizon and forces him to retreat to the caves, to safe spaces where he could live his life and create a meaningful world. *Work on Myth* describes a similar process in more detail. In this book, the first experience of the newborn species is the "absolutism of reality," the threat of a dangerous imposition of otherness defined by the fact "that man came close to not having control of the conditions of his existence and, what is more important, believed that he simply lacked control of them" (1985b, pp. 3–4).

The human strategy against such uneasiness is the action of naming, because "to equip the world with names means to divide up and classify the undivided, to make the intangible tangible though not yet comprehensible" (Blumenberg, 1985b, p. 42). That is the first step in the construction of significance which leads to the identification of interactions and relations between the realities which have been previously determined. Hesiod's myth of creation can be an example of such a process, as chaos becomes gradually defined by submitting the disorder under the name of different deities. Therefore, myth is the first creation of significance and, for that reason, "myth itself is a piece of high carat 'work of logos'" (Blumenberg, 1985b, p. 12) that turns chaos into cosmos. Blumenberg always highlighted the cosmic experience which underlies the Greek concept of reality, its reliance on the stability of the world and on its benevolence or, at least, indifference towards humankind. Hence, the passage from *mythos* to *logos* is not an epochal change, but the birth of a new way of explanation caused by the stiffness and excess of the mythical world

[20] These changes can be seen at least after 1971 with the publication of "Anthropologische Annäherung an die Aktualität der Rhetorik." This period was accurately described as an "anthropological turn" in Blumenberg's thought (Müller, 2005, p. 18).

which, however, stands on the shoulders of the cosmic experience produced by the myths.[21]

So, in *Work on Myth* the creation of significance is the first step towards the stabilization of a protective cultural shield which keeps the absolutism at bay. However, this absolutism may return when this fragile shield collapses. In *The Legitimacy of the Modern Ages* Blumenberg had already explained how Gnosticism, as a system which separates salvation and existence, was the symptom of the loss of confidence in the ancient order, an experience that resurfaced in the late Middle Ages with the *Deus absconditus* from nominalism and voluntarism (see Blumenberg, 1985a). This theological absolutism emphasized an unsurmountable breach between experience and expectation, the experience of the lack of control over reality which revealed the unfeasibility of the horizon of experience. As we can see, there is a regularity in the process of crisis and restoration of the horizon that was evidenced in the passage to Modern Ages. The uneasiness of the lifeworld caused by the breakdown of the protective shield is the experience of the loss of order caused by the existence of breaches that allow that threatening absolutism to seep through. This dissonance in human daily life triggers anthropological processes to restore coherence even though the resolution may vary in each instance.

Barbara Merker wrote that, in Blumenberg's understanding of history, "there are two extreme situations from which [...] we must and can move away. Lifeworld is Blumenberg's name for one of those limit values. Absolutism of reality is his title for the other" (1999, p. 68). These two poles are the limits of a historical tension that can explain the instability of human existence. Thus, historicity will be the name of the process of construction and dissolution of provisional orders, of surrogates of the primary lifeworld that are able to shelter humankind and provide him with a safe place from which to comfortably develop the labour of consciousness and reason. Thus, history becomes the response to an anthropological need that is intimately bonded with the structure and features of intentional consciousness.

Here the intentional skeleton of historicity becomes fully visible. The motivation underlying the succession of history is the necessity of the human being for a protective shield to live in. That is the motor of historical intentionality, to distance the threat of the absolutism of reality and the reason that makes changes take place. This occurs when the balance of the lifeworld fails to overcome its internal difficulties and causes the activation of the processes of restitution. As we can see, the importance of the phenomenological descriptions of consciousness in Blumenberg's historical understanding that we have advanced is decisive and definitively outlines the figure of his flow. At this point, the reach of the phenomenological influence becomes fully understandable.

[21] In this sense, Blumenberg understands Thales' famous sentence "everything is full of gods" as the reflection of the weariness of an "administered world" ossified with the pedantry of systems in crisis (Blumenberg, 1985b, p. 128).

6 The Rationality of History

This phenomenological historical scheme is not just a narrative model for the unification of the diverse and the everchanging succession of horizons through time. Rooted in anthropology, it provides a rationalization of history itself. Its course is driven by needs that must be addressed to ensure self-preservation. Therefore, the theoretical ideal, and even its later results such as formalized science and technology, are legitimate responses to the challenges posed by a threatening otherness. However, this rationalization is not the unfolding of a transcendental rationality but of a contingent one, which is a result of the problems posed by evolution. This implies a restriction of the power of reason and suggests that the end of history may not be a desirable goal as projected by idealist philosophy of history, as it is intertwined with our anthropological constitution. Blumenberg envisions the possibility of a loss of motivation caused by the final domestication of the threats of the absolutism of reality and describes this state using the physical idea of entropy; a final state of dissolution that may have unexpected consequences as it will render intentional consciousness useless. From the perspective of the later Blumenberg, this would imply the loss of a function for a human disposition developed during our evolution. The generalization of boredom, a "diagnosis with a great future" (Blumenberg, 2006, p. 703), is one of the signals of the proximity of such a state. Blumenberg describes boredom as "being out of oneself without reality" (2006, p. 702), as "poverty of intention" (2006, p. 707), a state that can be very harmful.[22]

These descriptions may seem very current and show us the validity of such an analysis to understand our present. However, his validation of history is disputed and polemic. In some ways, Blumenberg opposes his understanding of history to the cultural critics; his vindication of the theoretical ideal counters the complaints against the enlightenment of Adorno and Horkheimer, for example.[23] Therefore, his vision can be accused of overlooking the victims of a process that was neither peaceful nor fair. This can be a consequence of the position against critical theory of a work which, on the other hand, is not blind to violence[24] and provides some critical perspectives for an analysis of the probable future paths of humanity in time.

---

[22] Blumenberg briefly catalogs central events of the twentieth century that could be attributed to boredom: "Heinrich Mann sees the fall of the Weimar Republic arising from boredom. [...] Hannah Arendt sees Adolf Eichmann emerging from an 'abyss of boredom.' Finally, what happened in Paris in May 1968, which should be neutrally called 'the events' interpreted itself in the flyer 'France was bored'" (Blumenberg, 2006, p. 706). Josefa Ros Velasco (2022) has researched the topic of boredom from Blumenberg's understanding of the subject in a broad sense.

[23] For the polemical relationship between Hans Blumenberg and the *Frankfurter Schule* see Benlliure (2015) and García-Durán (2019).

[24] The concept of the absolutism of reality that we have analyzed in these lines is an example of Blumenberg's concern with violence. However, his most interesting writing on this topic is the aforementioned paper "Anthropologische Annäherung an die Aktualität der Rhetorik," which was first published in 1971 and then included in the compilation *Wirklichkeiten in denen wir leben* (Blumenberg, 1981).

For example, Blumenberg suggests the possibility of constructing a homeostatic virtual horizon that could fully protect humanity from reality a "definitive cavern" (see García-Durán, 2020, pp. 395–401) maintained by the possibilities provided by a spectacular society. This could be a logical outcome of the desire to distance oneself from the threatening reality facilitated by technological development and the virtual digital world. While it may appear to be a trivial sci-fi scenario, this possibility may not be far off in time, and its consequences could expose the risks inherent in human mechanisms for the creation of signification. Such a scenario may not be as desirable as it seems, as it would lead to the loss of the utility of the mechanisms of consciousness, as exposed earlier. However, these dystopian scenarios from Blumenberg's work have been largely overlooked in his reception (García-Durán, 2020). The implications of this intentional interpretation for understanding the procedures created to keep consciousness occupied and the risks involved in accurately considering the limits and possibilities of this expansive system are worth investigating. We may be prisoners in the cave, entertained by the succession of shadows and voices, but is the show trying to conceal the weakness of the walls on which it is projected? Blumenberg's theory does not offer a critical standpoint against a deceptive cultural order, but it does contain warnings and advice on the risks of the rise of absolute powers in such critical situations. It also provides insight into the anthropological significance of historical thought, as only through humanity's efforts to overcoming the threat of the absolutism of reality, has humanity been able to achieve the dignity that it has obtained. In that sense, as the German philosopher stated in the reception of the Kuno Fischer Prize in 1974, 'preserving history is preserving mankind' (Blumenberg, 1981).

References

Accarino, B. (1999). Nomadi e no. Antropogenesi e potenzialismo in Hans Blumenberg. In A. Borsari (Ed.), *Hans Blumenberg. Mito, metafora e modernità* (pp. 287–340). il Mulino.

Benlliure, R. (2015). Blumenberg y la teoría crítica. In F. Oncina Coves & P. García-Durán (Eds.), *Hans Blumenberg: Historia in/conceptual, antropología y modernidad* (pp. 205–222). Pre-Textos.

Blumenberg, H. (1962). Ordnungsschwund und Selbstbehauptung. Über Weltverstehen und Weltverhalten im Werden der technischen Epoche. In H. Kuhn & F. Wiedmann (Eds.), *Das Problem der Ordnung* (pp. 37–57). Hain.

Blumenberg, H. (1981). *Wirklichkeiten in denen wir leben*. Reclam.

Blumenberg, H. (1985a). *The legitimacy of the modern ages*. MIT Press.

Blumenberg, H. (1985b). *Work on myth*. MIT Press.

Blumenberg, H. (1986). *Lebenszeit und Weltzeit*. Suhrkamp.

Blumenberg, H. (1989). *Höhlenausgänge*. Suhrkamp.

Blumenberg, H. (2002). *Zu den Sachen und zurück*. Suhrkamp.

Blumenberg, H. (2006). *Beschreibung des Menschen*. Suhrkamp.

Blumenberg, H. (2010). *Theorie der Lebenswelt*. Suhrkamp.

Blumenberg, H. (2018). Phänomenologische Schriften (1981–1988). Suhrkamp.

Blumenberg, H. (2022). *Die ontologische Distanz. Eine Untersuchung zur Krisis der philosophischen Grundlagen der Neuzeit*. Habilitationsschrift (1950). Suhrkamp.

Buch, R., & Zambon, N. (Eds.). (2023). Hans Blumenberg und die Phänomenologie. *Phänomenologische Forschungen*, 2.
Carchia, G. (1999). Platonismo dell'immanenza. Fenomenologia e storia in Hans Blumenberg. In A. Borsari (Ed.), *Hans Blumenberg. Mito, metafora e modernità* (pp. 215–226). il Mulino.
Fragio Gistau, A. (2012). Hans Blumenberg: Las dificultades de la Ilustración a través de sus metáforas. *Anales del Seminario de Historia de la Filosofía, 29*(2), 65–96.
García-Durán, P. (2015). Un *eidos* de la historia. Elementos para una fenomenología de la historia. In F. Oncina Coves & P. García-Durán (Eds.), *Hans Blumenberg: Historia in/conceptual, antropología y modernidad* (pp. 223–239). Pre-Textos.
García-Durán, P. (2017). *El camino filosófico de Hans Blumenberg*. Alfons el Magnànim.
García-Durán, P. (2019). El mito como concepto polémico. El trasfondo político del debate sobre el mito en 'Poetik und Hermeneutik'. *Conceptos Históricos, 5*(7), 128–155.
García-Durán, P. (2020). Imágenes del final del camino. Utopía e historia en el pensamiento de Hans Blumenberg. In J. D. Bares Partal & F. Oncina Coves (Eds.), *Utopías y ucronías. Una aproximación histórico-conceptual* (pp. 385–405). Edicions Bellaterra.
Hernández Marcos, M. (2015). Indigencia natural del hombre y modernidad en Blumenberg. Una respuesta a la insuficiencia del análisis de Husserl. In F. Oncina Coves & P. García-Durán (Eds.), *Hans Blumenberg: Historia in/conceptual, antropología y modernidad* (pp. 121–134). Pre-Textos.
Husserl, E. (1954). *Die Krisis der europäischen Wissenschaften und die transzendentale Phänomenologie. Husserliana* Bd. VI. Nijhoff.
Jauss, H. R. (1969). *Nachnahmung und Illusion, Poetik und Hermeneutik I*. Fink.
Landgrebe, L. (1963). *Der Weg der Phänomenologie*. Gütersloh.
Lederle, S. (2021). *Endlichkeit und Metapher. Studien zu Hans Blumenberg und Eugen Fink*. Königshausen u Neumann.
Merker, B. (1999). Bedürfnis nach Bedeutsamkeit. Zwischen Lebenswelt und Absolutismus der Wirklichkeit. In F. J. Wetz & H. Timm (Eds.), *Die Kunst des Überlebens. Nachdenken über Hans Blumenberg* (pp. 68–98). Suhrkamp.
Merker, B. (2009). Phänomenologische Reflexion und pragmatistische Expression. Zwei Metaphern und Methoden der Philosophie. In A. Haverkamp & D. Mende (Eds.), *Metaphorologie. Zur Praxis von Theorie* (pp. 153–180). Suhrkamp.
Monod, J. C. (2002). *La querelle de la sécularisation. De Hegel à Blumenberg*. Vrin.
Müller, O. (2005). *Sorge um die Vernunft. Hans Blumenbergs phänomenologische Anthropologie*. Mentis.
Philippi, M. (2019). Das selbstverständliche Verstehen. Zu Blumenbergs Auseinandersetzung mit Edmund Husserl. In A. Fragio, M. Philippi, & J. Ros Velasco (Eds.), *Metaphorologie, Anthropologie, Phänomenologie. Neue Forschungen zum Nachlass Hans Blumenbergs* (pp. 146–165). Alber.
Ros Velasco, J. (2022). *La enfermedad del aburrimiento*. Alianza Editorial.
Stoellger, P. (2000). *Metapher und Lebenswelt: Hans Blumenbergs Metaphorologie als Lebenswelthermeneutik und ihr religionsphänomenologischer Horizont*. Mohr.
Welter, R. (1986). *Der Begriff der Lebenswelt. Theorien vortheoretischer Erfahrungswelt*. Fink.
Zambon, N. (2019). Zwischen Husserl und Heidegger. Der Weg zu einer Beschreibung des Menschen. In A. Fragio, M. Philippi, & J. Ros Velasco (Eds.), *Metaphorologie, Anthropologie, Phänomenologie. Neue Forschungen zum Nachlass Hans Blumenbergs* (pp. 73–90). Alber.
Zowislo, M. (2009). The existential overcoming of phenomenology in Hans Blumenberg's philosophy of life and myth. In A. T. Tymieniecka (Ed.), *Phenomenology and existentialism in the twentieth century (Analecta Husserliana, vol. 103)* (pp. 309–321). Springer.

Pedro García-Durán holds a Ph.D. in Philosophy from the University of Valencia with a thesis on the phenomenology of Hans Blumenberg. He has been an associate professor at the Universitat

Jaume I (Castellón de la Plana) and the Universitat de València. He is a member of the research group "Conceptual History and Criticism of Modernity" at the latter institution. He has published the first monograph in Spanish on this same author, *El camino filosófico de Hans Blumenberg: Fenomenología, historia y ser humano* (Institució Alfons el Magnànim, 2017), an author on whom he has also co-edited with Faustino Oncina Coves the volume *Hans Blumenberg: Historia in/conceptual, antropología y modernidad* (Pre-Textos, 2015). More recently, he has edited with the aforementioned Professor Oncina the collective book *Mitologías políticas* (Plaza y Valdés, 2024). He has published in numerous journals and collective works, both in Europe and Latin America, such as *Revista de Filosofía Aurora, Historia y Grafía, Conceptos Históricos or Prometeica.* His field of research also includes German conceptual history, as well as the relationship between politics and significance.

Lifeworld and Historicity: The Fate of Phenomenology

José Luis Villacañas Berlanga

Abstract On the occasion of the Spanish edition of *Home and Beyond: Generative Phenomenology after Husserl*, by Anthony J. Steinbock, this essay reflects on the dynamism of Husserl's phenomenology. Dominated by his aspiration to turn phenomenology into a self-sufficient philosophical science, Husserl was confronted with the inevitability of an originary unconscious dimension, the retentional zero that complicates self-conscious life with the impenetrability of passive syntheses. After analyzing the static and genetic dimension of phenomenology, the essay studies the meaning of the generative program that Husserl mentioned in some texts starting from *Cartesian Meditations*, reconstructed by Steinbock with a firm hand, I try to show that this generative program implies a new understanding of phylogenetics of consciousness. Husserl only announced it and the proposal of the structures of the familiar world and the foreign world do not seem sufficient to achieve it. My final proposal is that this program can only be deployed as something more than a proposal under the inspiration of the imposing work of Hans Blumenberg.

The Spanish translation of Anthony J. Steinbock's book *Home and Beyond: Generative Phenomenology after Husserl* (1995) is a philosophical event that should not go unnoticed. Moreover, the quality of the Spanish version verges on perfection, for which we must thank the work of Rocío Garcés and Andrés Alonso (2022). However, the best way to pay tribute to the book is simply to study it and explore the possible ways in to deploy its perspectives.

I will begin with a somewhat dogmatic tenet that will serve as a starting point. All the elements that make it possible to discuss this book involve the problem of the lifeworld (*Lebenswelt*). That is because this concept is arrived at from the initial

English translation by Josefa Ros Velasco.

J. L. V. Berlanga (✉)
Universidad Complutense de Madrid, Madrid, Spain
e-mail: jlvillac@ucm.es

A. Fragio et al. (eds.), *Hans Blumenberg's History and Philosophy of Science*, Studies in History and Philosophy of Science 63,
https://doi.org/10.1007/978-3-031-78251-0_5

reflection on the conditions of a transcendental philosophy. Indeed, it arises from the will to discover the fullness of the conditions of phenomenology in order to guarantee its full autonomy as a theory. I intend to devote the first section of this chapter to this question. It will consist of several parts: the first will give a brief genealogy of the concept; the second will deal with its specific functionality in Husserl; the third will briefly describe the basic levels of phenomenology in relation to it; the fourth will explain what I consider to be the central problem of these initial levels; the fifth will set out the programme of a generative phenomenology; and in the sixth, I will show how this programme can only develop through Blumenberg's insights. The fate of phenomenology will, in my view, depend on these insights.

1 Facticity

We are familiar with the history and prehistory of this concept of the lifeworld, so an outline of it will be sufficient here.[1] Its origin lies in the revision of neo-Kantianism carried out by positivism. Since Kant, some ultimate condition of facticity had always been arrived at as a complement to transcendental discourse. Kant's *Faktum der Vernunft* marked the inner limits of transcendental analytics, which strove to understand the conditions of possibility of a reason that already existed. At the same time, this limit was the starting point of transcendental discourse. By means of this expedient, classical transcendental discourse demonstrated its impossibility of becoming an absolute discourse. It could not integrate all its own conditions of possibility; the facticity of what it analysed was prior to it. At the basis of the analysis stood the *Faktum* of that of whose conditions of possibility had to be accounted for. Critical discourse never assumed that ascertaining the conditions of possibility of reason implied at the same time producing the facticity of reason; that required a history of reason as a complement to its transcendental analysis. Fichte's philosophy, in contrast, set out the transcendental analysis of reason as if it were at the same time the process of a *Setzung* of reason. In Kant this is not the case. Analysis and *Setzung* are two different structures. The existential process of *Setzung* is not derived from transcendental analysis. In Kant, the *Faktum,* which is always presented by a *Setzung* of its own, was both the limit and the starting point of transcendental discourse. It only explained the conditions of possibility of knowing something already existing in its facticity. Existence thus has an ontological priority and independence. In Kant it is linked to transcendental aesthetics.

When the neo-Kantians deployed Kant's thought at the end of the nineteenth century, they replaced this *Faktum* of reason with the *Faktum* of science. The latter existed, it preceded philosophical self-consciousness and it was a matter of seeing how it was possible. Philosophical self-consciousness had to show both the

[1] For the prehistory of the *Lebenswelt,* see the introduction by the editor Rochus Sowa to volume 39 of Husserliana (Hua 39 [Husserl, 2008]), *Die Lebenswelt*.

conditions and the limits of science and thus display other facticities such as morality, aesthetics, religion, and politics. This scheme was maintained by Lukács in Heidelberg, and this approach was applied to aesthetics in a move that the neo-Kantian Max Weber always applauded.[2]

Of course, this path must be taken into account when we reach Husserl. However, the specific history of the term *Lebenswelt* comes from neo-positivism. This current took a step beyond neo-Kantianism and showed that the *Faktum* of science was not the original *Faktum*. Science always relied on other conditions of factuality. These came together in what Mach and Avenarius called *der menschliche Weltbegriff*—the concept of the human world. This consisted of a natural attitude to the world (*die natürliche Einstellung*) in which a human being was spontaneously established. This concept of the human world was natural (*die natürliche Weltbegriff*) and anchored in conventional, everyday life. It offered the set of assumptions that were at the basis of the formation of theoretical concepts. Mach and Avenarius, and later others such as Neurath, deployed significant arguments to show the continuity between the basic concepts of science and this natural attitude to the world; and Husserl was raised in this environment. For example, they linked the basic sense of temporality founded on the notion of *now* with the pulse by which humans feel their own heart beating.[3] In this way, they launched the idea of showing how theoretical categories pointed to basic vital conditions that were linked to the elements of life itself. In this way, a natural attitude as a subjective formation became associated with the objective correlate of the natural environment (*die natürliche Umwelt*), which biologists soon made fashionable.[4]

From very early on, Husserl made free use of these elements, and, from the beginnings of his philosophy (*Idee*, Hua 2, 17 [1950b]), he understood that this natural attitude implied a direct orientation towards things. This orientation was completely alien to scepticism. Its main characteristic was the *Generalthesis* that the world really exists. Its ancestor could be seen in Hume's *Belief*, which Jacobi had introduced into German philosophy, and this attitude as such was considered to offer a *Dingwahrnehmung*. In *The Basic Problems of Phenomenology* (published 1910), Husserl stated

> It could be shown that philosophical interests of the highest dignity require a complete and comprehensive description of the so-called 'natural concept of the world': that of the natural attitude; and furthermore, that an adequate and profound description of this kind is not

---

[2] At the very beginning of the *Heidelberger Ästhetik (1916–1918)*, Lukács says: "There are works of art, how are they possible?" (1975). And at the beginning of the *Heidelberger Philosophie der Kunst (1912–1914)*, Lukács also establishes that "in this work the factum of aesthetics is analyzed and from it, with the Kantian way of questioning, the systematic construction of aesthetics is rehearsed" (1974). For Max Weber, see Karädi and Fekete (1982, esp. p. 362).

[3] For the relations of Avenarius and Mach with Husserl, apart from the introduction to Hua 39, see Blumenberg (2006, esp. pp. 50, 84, 149–153).

[4] It was in this sense, as an organic world, that Max Scheler used his concept of *Lebewelt*, even before Husserl. Blumenberg (2010, see. p. 21), remembers that he did so in *Zur Phänomenologie und theorie der Sympathiegefühle* (1913, see p. 27 [1973a]).

easily achieved, but on the contrary requires reflections of enormous difficulty (Hua 13, 124–125 [1973a]).

Thus, phenomenology was initially the attempt to make the natural world fully self-conscious. The *epoché* is a suspension with respect to the natural world. What is given to phenomenological theoretical consciousness must maintain a relation to the 'pregiven world.' When, in the texts from 1918 onwards, he begins to approach the problem in terms of *Lebenswelt*, the analysis is entitled *Auslegungen der vorgegebenen Welt.*[5] There is thus a continuity between the terms of his positivist predecessors and the problem of the *Lebenswelt* as the factual basis of phenomenology. Thus, the concept of *Lebenswelt* took over the function of all other preceding terms.

2 The Lifeworld as Conditioning Facticity

This pregiven world integrates the instance of ontological priority of classical transcendental analysis shared by phenomenology. As Steinbock says, "Husserl again emphasizes the departure for phenomenology as ontology" (1995, p. 80, 2022, p. 132). It is a question of what will later be called in *Krisis* (Hua 6, 51 [Husserl, 1954]) the "ontology of the lifeworld." Steinbock rightly places here what he calls the level of static phenomenology. His aspiration is to identify the eidetic structures that emerge from intentional games, the necessary relations of meaning that are imposed on us in their noemes, now as objective referents of fully conscious noetic acts. The aim of this analysis was by no means to close down the immediate senses, but to show their validity as conditions of science and theory. He thus attempted, as we see in Husserliana (Hua) 13, 196 (1973a), to link phenomenology with the salvation of the original senses themselves as foundational experiences more profound than the categories of the sciences. This would allow the preservation of an original philosophical experience. In fact, the subsequent observed crisis of European sciences occurred when this understanding of fundamental structures anchored in lifeworlds was lost, and post-Copernican science was endowed with what was seen as absolute autonomy. Husserl explained this in his delightful work "The Earth Does Not Move" (2006).

Related to this question of the relationship between phenomenology and natural attitude is the idea of going back to the things themselves. These things themselves are those of the natural attitude, and the aim is to preserve them in their originality. Steinbock clearly perceives that the concept of the lifeworld is central and multifaceted in Husserl. At any event, it always continues to play the dual function of the

[5] Husserl's first texts prior to the concept of the lifeworld speak of "the predated world as the original acquisition [*ursprünglicher Erwerbung*], in front of consciousness, which is defined as the realm of immanence. Of course also the world of culture [*Kulturwelt*] is considered as predated" (Hua 39, 19–34 [2008]).

facticity assumed by phenomenology while also offering the beginnings of a phenomenological approach. In this regard, we can say that the lifeworld is maintained as long as the ego is in the stream of life, "in the natural pursuit of life" (Hua 4, 375 [Husserl, 1952]). On this same page, we are told that the lifeworld is the natural world "in the attitude of natural life there" ["in der Einstellung des natürlichen Dahinlebens"] and that we are thus "living subjects in an open circle of other living subjects." This subjectivity is a lived experience of *Umwelt*. Of course, we have the primordial (*urtümlich*) attitude here. Here the subject is always the psychological ego. However, when the ego withdraws from the stream of life, from 'executivity,' to use Ortega y Gasset's term, then it becomes the transcendental ego which realises the *epoché* and raises its noetic acts to the condition of all its noematic contents.[6] This is the ultimate reason why Husserl advocates a parallelism between the two egos and why he speaks of life and a transcendental life; one given over to consciousness, and the other to the series of self-consciousness insofar as it makes conscious the idea that all representations belong to an 'I' that is acting behind them.

In this way, the programme was imposed that the main aspiration of phenomenology was to render the lifeworld fully self-conscious. This is what characterised phenomenology as the science of trivialities, as Heidegger said back in 1919 (see von Wolzogen, n.d.). However, Husserl's aspiration was not to make problems of these trivialities, as would happen in Heidegger, but simply to show their constituent acts. In fact, the fundamental question that must be placed at the origin of the different levels of phenomenology is: why not simply live in the lifeworld? Why the obligation to describe it? Where does the need for self-consciousness come from? And even why go beyond description or static phenomenology? In short, why take the step to phenomenology and furthermore to a genetic phenomenology? Steinbock speaks of a first, static phenomenology, which is the one that takes the lifeworld as an assumption and tries to grasp its essentialities through a phenomenological reduction or *epoché*. However, Husserl wanted to respond to Nietzsche by showing that those structures that appeared as necessary in the lifeworld—spatiality, temporality, successions, retentions, protentions—were not mere habitualities or normalities. They were also legitimate. And for that he had to go beyond static phenomenology and show their structure of necessity.

3 Genetic Phenomenology

I think Husserl's answer to these questions would be that we come out of the lifeworld for logical reasons. Incoherences, associations, passive syntheses and deformations criss-cross the lifeworld with meanings that need to be refined. This is Bacon's problem. It is the need to move from pre-predicative to predicative attitudes; to move from presuppositions and assumptions to enunciative meanings; to

---

[6] For the relations between Ortega y Gasset and Husserl, see Villacañas Berlanga (2024).

move from habitual normality to logical normativity; or in Hegelian terms, to move from belief and certainty to truth. In short, theory still serves the lifeworld, but emerges from an idea of logical truth that seeks to refine and legitimise it. In this emergence, task and function validate the substantiality of the *Logos*. This serves the lifeworld insofar as it shows its profound duty, its normative logical being, beyond what is our normal habit. It is also the passage from the imagination to the rule of Kant's concept. The predicative *Logos* certainly introduces a new modality alien to the lifeworld. It recognizes the modalities of necessity, possibility, effectiveness, falsity, and truth in its compact facticity. Steinbock recognizes this difference correctly. The lifeworld has no modality. The *Logos* that is exercised in genetic phenomenology does.

Now, this argument cannot be made merely with static phenomenology. It cannot be satisfied by identifying eidetic forms that reveal what is implicit and implied in the natural attitude. Here, in the games of variations and noematic comparisons, Meinongnian essences emerge, which pass the filter of meaning, but not that of reference and truth. And the normatively full lifeworlds are those that resolve their internal tensions from truth. We thus have the utopia of a lifeworld philosophically assured in its legitimacy. This utopia would be a spontaneous and at the same time self-conscious world. We would thus achieve the goal of Platonism: we would have come out of the cave into the light and would return to the cave to give reality to all that has hitherto been shadows. This can only be achieved if we reconstruct normativity from self-consciousness, thus identifying the structure of necessity and truth in the lifeworld. This would be the meaning of genetic phenomenology. Its aim would be to unveil "the *foundational* relations of the possible experience of the worldly" or "the *foundational* relations of meaning" (Hua 15, 617 [Husserl, 1973b]). The thesis would be that there is no normativity or legitimacy without a self-conscious foundation. What happens to the institution of truth is what happens to any other institutional norm. It must invoke the foundation in order to identify its meaning. To do so, one must return to the dimension of "originarity as primordiality" (Hua 15, 617 [1973b]). Here we have the constitutive dimension of transcendental consciousness, which asks for the foundation of what in the lifeworld is offered to us as predicate. In these constitutive acts, transcendental consciousness operates in a "primordial present" ["urtümliche Gegenwart"] (Hua 15, 617 [1973b]). We seemed to be experiencing a Fichtean moment of phenomenology. It is a matter of appreciating in acts of self-consciousness the foundation by which the meaning and the references which are predicated in the natural attitude are originally and primordially laid.

Although Steinbock believes that genetic phenomenology is not transcendental phenomenology (I'm not sure about that), he appreciates its features very effectively, and at the same time presents its problems in a clear way. I believe that the question of transcendental phenomenology as to how the lifeworld becomes possible can only be solved by genetic phenomenology. Actually, I believe that transcendental phenomenology has two moments: the regressive and the genetic. The regressive or reflexive, called *Besinnung*, the anamnetic moment, must return to the instances of the foundations, and give way to the genetic or constitutive level. The transcendental ego drives this dual process, guided first by regressive and then by constitutive and genetic self-consciousness. After the dual process, it is certainly seen that the transcendental subject is the same subject of natural attitude, but torn from its "anonymous functioning," since the process is separated from its passive syntheses to be reconstructed in a fully self-conscious active synthesis with constitutive and normative value (Steinbock, 1995, p. 84, 2022, p. 138). The fact that this "world-constituting subject" cannot be separated from transcendental subjectivity is clear to Husserl (Steinbock, 1995, p. 84, 2022, p. 138).

The theory up to this point would be nothing but a monumental *amplificatio* of the cave myth. However, the difficulties in completing this level of phenomenology revolve around a fracture that forces us to break with the Platonic myth of the cave. The issue is that, in the end, we see a traveller who does not know when he has arrived in the world of light or, masterfully linking the myth of the cave with that of anamnesis, with a traveller who reaches the limits of what he can remember. The difficulty certainly lies in arriving at a sense of *Urstiftung* that can present itself in an original self-conscious present.[7] Here, of course, phenomenology encounters the same difficulties as psychoanalysis. Husserl, like Freud, was well aware of the limits of regressive-genetic phenomenology and of the necessity of an instance of the unconscious. Volume 42 of Husserliana (2014) is clearly devoted to this line of argument.

Let us now look at it in more detail. This is the problem of phenomenological archaeology highlighted by Steinbock. The text he proposes, from 1934, is fundamental to understand the problem I see here. It speaks of phenomenological archaeology as a layered excavation. The metaphor is the same that Freud uses for psychoanalysis, which he compares to the excavation of an ancient city, such as Rome or Troy. As in Freud's case, the constructions to be excavated 'are constitutively hidden.' There is no doubt that these hidden constructions correspond to what is visible as a result in the world of experience (*Erfahrungswelt*). Husserl speaks of their being 'found there.' Starting with the results found there, a retrospective path begins. With it, a regressive process reveals the *archai,* by which Husserl alludes to 'the sense of being of even the last of the singular creative operations.' But once the *archai* are reached, it is a matter of rising in 'the spirit.' If we ask how and how far, the answer is clear: by means of the 'self-evident unity' that the elements of validity reveal in their validity until the lifeworld is reconstructed in its purified being. We thus have a 'reconstruction.' Husserl speaks of a zigzag march. One goes back, and through this revelation of what is hidden one reconstructs with evidence what is given there in its validity (see *Späte Texte über Zeitkonstitution* [published 1934],

[7] For the difference between *Fundierung* and *Stiftung*, it is important the work of Sam Gault (2017). The former is the structure of lived experience. It is related to the instances of static phenomenology and discovers the links of necessity in the noematic contents. *Stiftung* is related to acts of synthesis and presupposes the unity of the ego with its temporal sequence of genesis of the life of consciousness that unifies the different founded *cogatita*. *Foundierung* does not mean the temporal consciousness of genesis. *Stiftung* institutes a habituality or tradition of conscious life, a permanent possession (see *Cartesian Meditations* §§18, 27). For further details, see Merleau-Ponty (1998).

2005, Hua/Mat, 8, 365–367 [Husserl, 1959]). This makes clear the dual movement: to x-ray the hidden layers of constituted meaning and to reconstitute them up to the ontic levels of validity; to reach the foundations of the submerged city of experience and reconstruct it in its validity.

But how can we be sure that we have discovered the ultimate foundations of something predestined whose original construction took place in a hidden way when we were not there? Evidently only by the idea that alongside this psychological 'I' that lives in the lifeworld, there is a transcendental 'I' capable of installing itself in the original present. We cannot pass from the regressive moment to the genetic moment without relying on the assumption that the totality of the operations that took place in the occult, those demonic passive syntheses, which clearly configure a psychological unconscious, are transparent to the transcendental ego. Where there are passive syntheses there must be active syntheses. But that would claim the parallelism of both series. However, the psychological ego has an unconscious, but not the transcendental ego. As long as we are not sure that the transcendental 'I' is situated in the original present of the *archai*, the *phänomenologische Rekonstruktion* that was supposed to follow archaeology cannot be symmetrical to it. This concerns the foundations of validity. Only if we arrive at the *urprüngliche Zeitung* of an original foundation of meaning can we speak of what was already there before as a mere phenomenal appearance. This would imply that the primary ontology of the lifeworld has to be displaced by a secondary, legitimate ontology, endowed with a *geltende Wirklichkeit* (Hua 7, 2 [1956]). This secondary ontology is the result of the mediation of genetic phenomenology.

But how do we know that we have reached the end of the excavation? Freud, faced with the same problem, replied that analysis is infinite and that one cannot reproduce the formation of the psychological ego, but only give clues as to the formation of the symptom, to its false relation to reality. Of course, this reduction of the reflexive programme depends on the recognition of the existence of the unconscious. Now, the lifeworld has something of the unconscious in it. This will be one of Blumenberg's criticisms. There is always an inexhaustible excess of the lifeworld over the descriptions of the theory. Therefore, not all of its content can pass into the reflective, regressive process, and thus begin the reconstructive process. Determining what an *arché* is in an inexhaustible process is problematic. But beyond that we have the temporal problem. As Steinbock says, "*A lifeworld ontology* serves as a provisional, albeit necessary, stage in a general phenomenological regressive analysis that, ultimately, is submitted to a transcendental clarification" (1995, p. 94, 2022, p. 151). But how does the provisional make it possible to arrive at an *arché*? How can operations of which there is no notice in the psychological self be used as evidence by the transcendental self with self-evident and active lucidity?

4 The Unconscious

In Appendix X of Volume 42 of Husserliana, from the end of 1931, entitled *Das Rätsel des Unbewussten (Sedimentierung)*, we read:

> The difficulties into which we are led by the apparent obviousness that retention, understood as passivity, leads us to a retentional zero. But is the hypothesis of a retentional zero tenable? I try to think through to the end, as a possibility, the conclusion of the effective retentional transformation into a zero of transformation, a zero that can nevertheless be nothing, but a being sedimented in an immobile stillness (Hua 42, 62 [Husserl, 2014]).

We have here the postulation of a proto-activity that nevertheless plays with a retentional zero. There is a sediment that is the fruit of a synthesis that is proto-active but which takes place under a retentional zero. This sediment is the one that can passively cross with the activity of synthesis and not be discovered. However much that proto-activity is accessible to the transcendental 'I,' its zero-retentional operations forge an inert sediment that for reflection must appear as something purely passive. But if this is so, sedimented operations without retention are part of the result that we see as the lifeworld. They offer no possibility of becoming a noematic object of self-consciousness, since we have no retention of them as noetic activities. The lifeworld cannot become transparent and fully legitimate as long as we do not penetrate into those operations of zero retention. But how can this be done?

Steinbock expresses the same idea, but without alluding to this text, when he says,

> What I take issue with is not Husserl's lucid portrayal of the constitutive features of an objective sense. Rather, it is his tacit assumption that the world can function like an objective sense and thus can be clarified like a thing or object (1995, p. 101, 2022, p. 161).

The argument, as is well known, is Heidegger's. We can never place ourselves fully above the lifeworld in such a way that it becomes an object. Freud's infinite analysis is reproduced here, when Husserl says that the natural or lifeworld is a "halo or horizon of indeterminate determinability" (*Ideen*, Hua 3, 160 [1950c]). Phenomenology implies a programme of unveiling "the infinite network of hidden foundations" ["verborgenen Fundierungen"] which rises up for the phenomenologist, and of which those who "naively and plainly live in it" are ignorant (*Krisis*, Hua 6, 153 [1954]). But at a certain point, the phenomenologist is also ignorant.

Certainly, this implies forgetting some formulations of phenomenology that allude to the fact that phenomenology must begin at the beginning without soil (*bodenlos*) in order to create its own soil (*Krisis*, Hua 6, 185 [1954]). In reality, the sediment produced by proto-operations with retentional zero is never dissolved. Husserl describes it as the "*beständige Umgebung der Nacht des Sedimentierten*" (Hua 42, 62 [2014]): the constant environment of the night of the sedimented. It does not become transparent. But this means that we cannot avoid the "supposed layering of layers upon layers in something like a fusion in the universal night" ["das Aufeinander sich angeblich schichten zu so etwas wie einem Verschmelzen in der allgemeinen Nacht"] (Hua 42, 62 [2014]). Thus, the facticity of the lifeworld cannot be turned into complete normativity because it cannot show the titles of the original foundation. Steinbock recognizes this in another way: "Husserl takes phenomenological pains to counter a totalising perspective that would place the philosopher 'over' the earth-ground" (1995, p. 113, 2022, p. 176). The earthly precedence of the constitution sinks into the universal night. Phenomenological analysis, like psychoanalysis, is endless. It incorporates a dimension of *Endlosigkeit*.

5 Generative Phenomenology

I think the greatest innovation in Steinbock's book is to have foregrounded the problem of generative phenomenology as Husserl's attempt to circumvent these problems. This level of phenomenology, different from the genetic level—something that Merleau-Ponty (1968) did not always notice—would be the culmination of Husserl's effort: thought in a continuous process of becoming, experimental, multidirectional. Husserl would not cancel the value of static phenomenology, nor of genetic phenomenology; but to solve the problems of these two instances it would be necessary to go a step further. In fact, Husserl had been feeling this need since the fifth of the *Cartesian Meditations*. In §61 of this work, Husserl says that the 'genetic problems' had already been incorporated into the phenomenological work. Here everything operated according to transcendental phenomenology. However, he said that this did not touch on the 'generative problems.' These concerned birth and death. Certainly, they were decisive for the operability of the sedimented lifeworld, for they guaranteed the suturing of the limits of the psychological self and its incorporation into lifeworlds. But to these problems he linked the 'nexus of the generation of animality' (*Generationszusammenhang der Animalität*). These were problems of a higher nature because they involved 'a work of making the lower strata explicit,' the strata the *archai* of genetic phenomenology were presumably not sure to identify. Husserl thought that such work could not be planned and initiated for a long time. However, they were fundamental problems and in touch with tradition (CM, Hua 1, 169 [1950a]). Here the problem of the lifeworld was again invoked, for these future works concerned "a structural articulation of the primordial sphere in which we find a world, a primordial world" (CM, Hua 1, 169 [1950a]). There is reason to think that the reference here is to those strata and sedimentations of zero retention level. One sees a little when Husserl says that they would help "to clarify the origin of this primordial nature and of the primordial psychophysical [*leibseeliches*] unities" (CM, Hua 1, 169 [1950a]). Husserl calls these dimensions "immanent transcendences" (CM, Hua 1, 170 [1950a]), a good name for the last sedimentation. He has never defined the field of generative phenomenology as the culmination of his "gradual phenomenological system" (*Stufensystem*, CM, Hua 1, 170 [1950a]) better than here.[8] What is specific about this generative level is that it is involved in the need to explain the "psychological origin of representations" ("Probleme des psychologischen Ursprungs der Raumvorstellung, der Zeitvortellung, der Dingvorstellung," CM, Hua 1, 170 [1950a]) "within the framework of the universal nexus of the psychic constitution" ("Probleme des psychologischen Ursprungs der Raumvorstellung, der Zeitvortellung, der Dingvorstellung," CM, Hua 1, 170 [1950a]). This offered a "totally new configuration of psychology" that, although free of all psychophysical elements, would carry out a Copernican revolution that would definitively clarify the oft-promised key to the parallelism of intentional psychology and transcendental subjectivity ("Probleme des psychologischen Ursprungs der Raumvorstellung, der Zeitvortellung, der Dingvorstellung," CM, Hua 1, 170 [1950a]), for the work concerned these organic and psychic dimensions, nature and the psyche. That is why he argued that he would generate the field of "pure psychology" (CM, Hua 1, 170 [1950a]), of a "psyche in general" ("einer Seele überhaupt," CM, Hua 1, 171 [1950a]), which could not be explained without clarifying "the primordial psychophysical units" (CM, Hua 1, 169 [1950a]). Perhaps this point seriously compromises the pure dimension of this psychology. This study was to be undertaken through certain previously given threads. One of them is that of the familiar and the strange (*Fremdheit*) as dimensions of the primordial lifeworld. It was important to realize that in doing so, one was dealing with the realm of the human, self-endowed with a "own and mere sensitivity" ["meiner eigenen *blossen Sinnlichleit*"] (CM, Hua 1, 171 [1950a]). Husserl envisaged on this generative level a study of the constitution of the primordial lifeworld as a world of realities, which would imply a "complete phenomenology of nature" (CM, Hua 1, 172 [1950a]). In short, I do not believe that the generative horizon can be deployed without appealing to the natural dimension of animal psychophysical units and the difference between the animal psyche and human 'normality.' This difference cannot be established without psychophysical studies. Husserl himself recognizes this when he says: "The psychophysical fact of coming into the world leads to the problem of the bodily development of the individual (purely biological) and to that of phylogenesis, which in turn finds its parallel in a psychological phylogenesis" (CM, Hua 1, 168 [1950a]). But upon arriving here, Husserl disperses towards the question of monadology and makes humans and animals self-objectifications of monads.

We need not recall the unfinished nature of this Fifth Meditation, and with it the collapse of phenomenology as a movement. Fink did not prove capable of picking

[8] In fact, a little earlier he had said: "Nur dass damit [mit den genetischen Probleme] freilich noch die oben bezeichneten generativen Probleme von Geburt und Tod und Generationszusammenhang der Animalität nicht berührt sind, die offenbar einer höheren Dimension angehören und eine so ungeheure auslegende Arbeit der unteren Sphären voraussetzen, dass sie noch lange nicht zu Arbeitsproblemen warden können. Doch innerhalb der Arbeitsphäre seien hier noch gewaltige Problemgebiete genauer angedeutet (und zwar sowohl als statische wie als genetische Probleme)" (CM, Hua 1, 169 [1950a]).

up the threads that Husserl had left behind, and this was proved by Bruzina (2004).[9] Nevertheless, Husserl included in this meditation a sense of the direction of his thought. It was clearly leading to a renewal of metaphysics, and that is why Heidegger (1983) addressed himself to it in his 1928–1929 course on the fundamental concepts of metaphysics. This new metaphysics had to move away first from speculative excesses, from the old philosophy, and enter into what he called "a systematic explanation of the essential structure of human existence" (CM, Hua 1, 165 [1950a]). This would be done from the constitution of the primordial or original lifeworld through forms of doing and suffering that were to remain valid in the layers of meaning of the secondary world (CM, Hua 1, 162 [1950a]).

The most important part came later. For within the very idea of nature would be included an infinitely open realm embracing "an open multiplicity of humans (and in more general terms of animalia) distributed we do not know how in infinite space as subjects of a possible reciprocal community" ["Die offen endliche Natur selbst wird dann zu einer solchen, die auch in offener Mannigfaltigkeit unbekannt wie sic him unendlichen Raume verteilende Menschen, allgemeiner Animalien in sich fast als subjekte möglicher Wechselgemeinschaft"] (CM, Hua 1, 158 [1950a]).[10] This made the sense of the Earth as an ark concrete and the earlier reference to the generative connection with animality. All this would be studied at the generative level, a level from which the human-animal relationship could not be separated. That is why he said: "among the problems of anomalies there is also the problem of animality and that of the gradation of animals into superior and inferior" (CM, Hua 1, 154 [1950a]). We can leave aside the fact that for genetic phenomenology we would arrive at the evaluation of animality as an anomaly of humanity. From the point of view of generative phenomenology, it was only a matter of differences between their lifeworld, which, according to Husserl, required a much deeper phenomenological explanation. When we approach Plessner's work from these years, his theory of community in *Die Grenzen der Gemeinschaft, Eine Kritik des sozialen Radikalismus* (published 1924 [2002]), the phenomenological difference between the animal world and the human world in *Die Stufen des organischen und der Mensch, Einleitung in die philosophische Anthropologie* (published 1928 [1975]), or his theory of the difference between one's own and another's that articulates all political and cultural forms, in *Macht und menschliche Natur, Ein Versuch zur Anthropologie der geschichtlichen Weltansicht* (1931), we realize that, even without having this concept of generative phenomenology, he was addressing the problems that worried Husserl.[11]

For Husserl, it was obvious that this generative analysis had to be projected onto the horizon of a community and that is why he spoke in 1929, shortly after *Meditations*, of a "*tranzendetale Soziologie*" (Manuscript C8, 11a, Hua/Mat 8, 165

[9] I would like to draw attention to the relevance of Nietzsche in this stage of phenomenology (see, esp. p. 341ss.).

[10] Due to the relevance of the text, I note the original version in German.

[11] For an exceptionally instructive exposition of Plessner's work, see Joachim Fischer (2008).

[1959]). Steinbock, of course, makes use of these texts. Here, the level of community capable of constituting an original lifeworld implied the concretization at the *Stamm* level which only reached generative density through tradition. This gave us a concretization of the sedimentation of meaning which had been constituted by proto-activities of zero retentional level but which were nevertheless valid. Husserl was surprised at the wealth of approaches he was illuminating, leading him to the problem of the original inheritance of the generational transfer of meaning, etc. Finally, he realized that this gave him access to the problem of *Gechichtlichkeit*. When he raised the problem again in the *Nachlass* of the *Manuscripts of the Phenomenology of Intersubjectivity*, shortly after 1929, he left us a fundamental text. It is included in Husserliana 15, 138n2 (1973b), and states:

> The constitution of a lasting co-humanity [*Mit-Menschlichkeit*], of a general sociability [*allgemeine Gesellschaftlichkeit*] in a correlative relation to a practical surrounding world [*praktischer Umwelt*] can be approached in an abstract way before generation, so that an abstract historicity [*abstrakte Historizität*] is included there because of the kind of temporalization of this practical surrounding world as personally meaningful. But if generation comes into play, then this step forward in specificity is also a concretization [*Konkretisierung*] of the permanent co-humanities [*Mit-menschlichkeiten*] (mother, father, child) and with it we have a more concrete, generatively formed temporalization as well as a historical surrounding world.

Steinbock has delved into these texts and for this we owe him gratitude. He also reminded us that this orientation towards a concrete historicity can only be achieved through a *normale Form mythischer Stammvölker* (Hua 15, 432n1 [1973b]). This was the path towards a 'primordial constitution' (*Ur-konstitution*) that was to illuminate the superimposition of sedimented strata. Here the animal problem was raised again insofar as they could not live historically in a generative world (Hua 15, 160n1, 181 [1973b]), in a historical future that had its basis in the evolutionary creativity of the human community as specifically open.

Husserl did not show whether generative phenomenology solved the problem of the sediments not illuminated by genetic phenomenology. At any event, everything points to the fact that, like Freud, he was thinking of clarifying these hidden layers of zero retention, these unconscious and therefore necessarily abstract elements, from the specificities of the original worlds of communities constituted in an open historicity and capable of forging original worlds in the future. Animals did not have a place there. He did not show how this historical openness could lead to the clarifications of nature and the psyche, the forms of transmission of meaning that constitute the psychic ego, in order to arrive at the foundation of theory as a self-revelation of the transcendental ego. These indications remained up in the air. However, they were convergent with those that Cassirer, making use of the neo-Kantian analyses of subjectivity, had deployed in *Sprache und Mythos* (1925) and in the *Philosophy of Symbolic Forms* (2020).[12] We know to what extent Heidegger took other paths. Fink gave up and left the Fifth Meditation undeveloped. But we must remember that generative phenomenology started from the demands of concreteness of the

[12] For a study on this issue with additional bibliography, see Villacañas Berlanga (2021).

analyses of the original lifeworlds and that this was the demand and the common thread that had already emerged in 1919 when he wrote in the Appendix "Die Schichtungen der vorgegeben Umwelt" to *Natur und Geist* that "our analyses of the original lifeworld [*ursprünglichen Lebenswelt*] as a common human environment world [*als einer gemeinsamen menschlichen Umwelt*] required greater precision" (Hua 25, 327 [1987]).

6 Blumenberg's Suggestion

Why were the concrete lifeworlds from the generative point of view historical? Why were they open worlds? Why were they subject to the creativity of the generations? Why were these questions linked to the formation of the psyche in general? Why were the familiar worlds and the strange worlds not fixed in their relationship, but subject to a precise and open dialectic? On these specific questions we are not told anything. If generativity 'deals with historical and intersubjective becoming,' Husserl did no more than define the programme and show what role it could have in the realm of a phenomenology of levels. Steinbock acknowledges that all this could be part of a "self-explication of phenomenology itself" (1995, p. 260, 2022, p. 379) and assumes that Husserl did not finalise it, nor is it contained in Husserl. I understand these confessions as a desire to develop this programme. "Participating in this process," he tells us, "is not only the generation of phenomenology, but the phenomenology of generation" (Steinbock, 1995, p. 270, 2022, p. 391). In short, the fate of phenomenology depends on the capacity to deploy generative phenomenology. It should show how phenomenology itself was possible and, in this sense, make the generation of the fundamental elements of phenomenology legitimate and self-conscious. In doing so, phenomenology would integrate the totality of its own assumptions and its complete autonomy. This could only mean one thing: to explain the conditions of possibility of the transcendental ego as capable of self-conscious operations of full retention. Thus, generative phenomenology would be a recursive movement that would show the presuppositions of descriptive and genetic phenomenology. It would solve the problems left open in them, as we have seen.

Steinbock is right to invoke Merleau-Ponty as the true follower of this programme, which he developed possibly with some influence from Plessner.[13] My last point is to introduce the figure of Blumenberg right at this systematic moment. I do so with hesitation and caution, because Blumenberg himself is not a systematic thinker. Moreover, his research on these problems was not developed either in full studies on Husserl or in convincing and expressive developments of his own, for his analyses are generally to be found in posthumous writings. But they are

[13] On the last page of the prologue to the second edition of his book *Die Stufen des organischen, Einleitung in die Philosophische Anthropologie* (1975), Helmuth Plessner himself recognizes the surprising coincidences between his formulations and those of Merleau-Ponty, so he wonders if he has not read his book after all.

nevertheless extraordinarily rich in detail and can be a source of sound arguments. Blumenberg did not shy away from the demands of connecting with paleoanthropological, biological and psychological studies, as Husserl demanded. In doing so, he broke with Heidegger's idea that only the human is world-rich. Lifeworlds are as plural as living beings. The lifeworld was the fruit of successful selection and adaptation, and in it are structured the sediments we have seen in Husserl. One of the requirements for the crystallization of a lifeworld is that it has been maintained for a long time (Blumenberg, 2010). That is the measure of success. They have no organic constitution, nor do they constitute a merely representational framework. At any event, they stabilise the relationship of a living being to its *Umwelt* and of the living being to itself. Of course, they are forms of *ethos* with an organic basis and representational efficacy as ways of stabilizing the relationship with the *Umwelt*. They constitute long-term epochs as crystallizations of self-preservation. A lifeworld implies that a living being has arrived there.

This means that the whole lifeworld is inertial. Blumenberg says that it is attached "to the inertia of existence" ["tendenzielle Bestandsträgheit"] (2010, p. 103).[14] The self-preservation of the living being is linked to the maintenance of its lifeworld. For Blumenberg, self-preservation and world of life is "the limit concept of the permanent tendency in every world to the understood as implicit, and the limit concept of the initial value of an experience of the world in general" ["Die Lebenswelt […] als den Grenzbegriff der in jeder Welt bestehenden Tendenz zur Selbstverständlichkeit und als den Grenzbegriff des Ausgangswertes einer Welterfahrung überhaupt"] (2010, p. 65). As such, it is what does not feel the need to innovate and therefore does not consider alternatives. In reality, it does not consider alternatives but sinks into "pre-modality" (Blumenberg, 2010, p. 105: Prämodalität), which explains its great capacity to defend itself and resist in its naturalization. Everything in it is taken for granted and self-evident; what in the human world appears as prejudices and beliefs that do not need to be corrected each time but generate habits and *ethos*. They are thus a "sediment of an extremely selective repertoire of experience" ["Im Gegenteil beruhen die lebensweltlich gesicherten Anteile an der Welt, in der wir leben, gerade darauf, dass sie Sedimenten eines hochgradig selektiven Erfahrungsbestanden sind"] (Blumenberg, 2010, p. 63). It is of course enriched "at the pace of generations" ["Die Lebenswelt muss nicht statisch sein, wenn ihre Veränderungen im Zeitmass generationsbezogen sind"] (Blumenberg, 2010, p. 100). It is the space for the passivity of consciousness, for a passive signification free of interference. Passive synthesis is here the origin. Therefore, it makes the explanation of philosophy more difficult, not easier. All animals have a lifeworld and have not needed to open it to transformations. Their inertia can be broken by transformations of the environment, and these can be overcome if new forms of behaviour break through by random variations of biological innovations. But the lifeworld can only be destabilised from the outside (Blumenberg, 2010, see esp. pp. 16, 102). In

[14] In another original passage: "Auf der hochgradigen Verteidigungsfähigkeit des Lebenswelt […] beruht ihre geschichtliche Funktion" ["The historical function of the lifeworld […] is based on its great capacity to defend itself"] (Blumenberg, 2010, p. 104).

other words, lifeworlds can change by evolving, but that is something different from historicity. It implies openness of the lifeworld and the capacity to maintain accumulated transfers of meaning between two successive worlds. How this can be known implies the elaboration of a concept of reality about what has been lost.

Blumenberg's phenomenology implies that in the lifeworld no motives arise that lead to its self-annihilation. They are not subject to self-criticism. The over-understood is not identified as such. This means that for Blumenberg the temporality of the lifeworld is specifically anti-historical. In fact, Blumenberg says that the lifeworld is "the starting situation *par excellence* which must be assumed to be prior to all history" (2010, p. 91). This is the great finding about Husserl—that the concrete history of generative phenomenology begins with the loss of a lifeworld. A history is always relative to a lost lifeworld. That is why loss is a condition of possibility of history. The problem of how we know something lost is precisely the historical problem, and hence loss, historicity and the method of knowing what we cannot witness go hand-in-hand. Here we enter into a problem that Husserl did not raise. It implied recognizing that phenomenology cannot be autonomous because loss cannot be synthesised in an original present. It is not that it is actually unconscious; it is that everything is generated from a situation in which we were never there. Although the above basic elements of the features of the lifeworld are in Husserl, this point does not appear in the Husserlian tradition. The lifeworld is a necessary condition of history, but not a sufficient one. For historicity, its fracture is required, and the loss must be integrated and reconstructed as the prior situation to a subsequent situation. It arises from the question: how did we get here? In short, if we link the argument to the above, generative phenomenology needs to be concretised in the phenomenology of the fractures of the lifeworld and its forms of projection and retention in the subsequent situation. The concrete history, announced by generative phenomenology, implies the history of the ruptures of the lifeworld contained and saved or reconstructed in a subsequent situation, assuming that there is no presence of *archai*. Only there can the conditions be given to explain the emergence of self-consciousness and thus find a subsequent generative phenomenology.

However, the break from the lifeworld and the passage to self-consciousness is not due to logical reasons. It is not about the coherence of representations. While it is true that the world of representations is fragile, in the lifeworld they are selectively self-correcting. There are no logical reasons for leaving the lifeworld. Phenomenology, insofar as it is guided by self-consciousness, has no theoretical premises that are self-grounding. One does not leave the lifeworld by decision, nor does any latent *Logos* that longs to become self-conscious press one to do so. One leaves the lifeworld by a crisis provoked from the outside. The way in which a new one is constructed is what has to be explained generatively. Phenomenology thus has anthropological and historical premises, but not logical ones. It does not throw itself into theory by a decision. It only leaves a lifeworld when it is no longer possible to maintain its inertia. If the human being has history, it is because of this loss of a lifeworld. However, not every loss of a lifeworld makes it possible to reconstruct the world from which one left and thus maintain the generational continuity necessary for the possibility of a consciousness of the past. Not every lifeworld has the trait of historicity, and this must be explained. The fulfilment of these conditions is something that depends on certain circumstances and is not subject to any necessity. It is something contingent that had to be generated by the parties themselves. The beginning of phenomenology, or of the attitudes that will eventually lead to it, essentially self-consciousness, is a history and needs a crisis and a reconstruction. Without this prior sense of life, the historical sense of event and novelty could not be assumed. But the fact that the crisis of the lifeworld does not mean extinction or evolution can only be explained by the specific and concrete learning experience of the beings who achieved it. The way to explain the reconstruction of the world of loss after loss is the history of the emergence of reason. Human beings learned and did it.

In doing so, they constituted themselves as historical beings and human beings. It is in this constitution as human beings that the original and the primordial achieve their most precise meaning. But if it is generative then it must have a starting point and be understood as a concrete temporalisation. The consequence is that what may eventually become a descriptive, genetic or generative phenomenology should include historical self-consciousness. Thus, Blumenberg could be giving specificity to the programme of generative phenomenology.

In fact, Blumenberg (2010) claims that the subject is subjective only when his world has ceased to be stable, and then forges active syntheses. We can only consider that one has a motive for theory when in effect the return "to the lifeworld once lost can only be a theoretical measure" (Blumenberg, 2010, p. 74). Thus, theory would be a way of achieving self-preservation by other means, specifically representational, not biological; learning, not evolutionary. The motivation for self-consciousness and intentionality can only come from the aspiration for self-preservation by refinement of representations and can only be motivated by the very force of life; and it can only come into practice when all biological means of adaptation have collapsed with the loss of the previous lifeworld. Without explaining the rupture of the original lifeworld and the impossibility of recomposing it by biological means, phenomenology will not become aware of the history by which what is to be its next fundamental tool arises: self-consciousness.

The problem is that Husserl could not identify in its specificity that part beyond the lifeworld which subjects it to crisis, nor could he therefore reconstruct its loss. For him, the passage to active synthesis is a requirement of internal logical coherence. And this motivation cannot be applied to generativity, because concrete history lacks internal logic. It is a future. Although there is certainly no other way of adjustment than the refinement of representations, in the original phases of generativity no theoretical interests as such can be deposited. There everything must be at the service of the self-preservation of life. The fact that this implied an anthropology was clear to Lévy-Bruhl when he dedicated the German edition of his *La metalité primitive* to Husserl (1929), after listening to his *Cartesian Meditations* in Paris in

1928.[15] Of course, by employing the support of anthropology, Blumenberg kept alive his anti-Gnostic aspiration, and tried to separate phenomenology from its Platonic ancestor, by showing the real conditions of theory, far from the unworldly messianism of the *Logos*. His critique of Husserl was that Husserl's original foundation was *presented* as an original decision in favour of theory. In this sense Blumenberg will say that "the arbitrariness of the archaic act places at the root of history a *decisionism* that removes it from all theoretical disquisition" ["Die Willkür des urzeitlichen Aktes setzt einen Dezisionismus an die Wurzel der Geschichte, der ihn jeder theoretischen Erörterung entzieht"] (2010, p. 18), injunction which bound humanity to a task that developed subjectivity through all human bodies aspiring to truth. Every historical epoch measured its legitimacy in relation to its fidelity to this original foundation which, according to Nietzsche, had to be secured in order for us to remain within the identity of Western history. This had as its essential backbone the history of theory. The question that remained unasked was this: why does life not remain in the lifeworld? Why is there a passage to self-consciousness, which is a competence not necessary in that world? Why is a procedure for the improvement of representational performance set in motion without explicit theoretical interests? What is the prehistory of theory?

What is decisive in Blumenberg is not the relativisation of the European history of theory, but two other points. First, the questioning of the ultimate condition of theory, which is self-consciousness. Thus, theory was one of the extreme forms of self-consciousness, the most intense and the strangest. It was not two simultaneous moments of generativity. One, self-consciousness, was related to achieved anthropogenesis. The other, theory, was related to the Greeks. Humanity could not be confused with Greece. Second, it was necessary to describe the rupture of the original lifeworld with respect to the being that would become the human being in order to understand the generation of self-consciousness. In this way, generation had to be considered in concrete terms on the basis of the animal connection, something that Husserl had intuited. For the lifeworld prior to the emergence of self-consciousness could only be an animal lifeworld. All the specificity of generative phenomenology came together here. The consequence implied a generalisation of the category of the lifeworld to all living beings and thus to identify which lifeworld was lost to give way to one in which self-consciousness was a necessary element for self-preservation and how it was achieved, with its difference between familiar and alien world. The question of why we are like this inevitably leads to the question of what we lost and why, and what we gained and how.

In short, Blumenberg asks about 'the essence of what has been initiated there.' This implies a complete synthesis of phenomenology and anthropology. And this is the central point of his posthumous book *Beschreibung des Menschen* (2006). And that is why my thesis is that the programme of a generative phenomenology, as Steinbock has defined it, can only develop through a new conjunction of

[15] Husserl's Letter to Lévy-Bruhl ist included in Hua. Dok. III (1999). See also Blumenberg (2010, esp. p. 71).

phenomenology and anthropology. The fate of what is still the program of phenomenology, namely a theory of human reason, depends on this conjunction. It is a pity that Blumenberg did not fully develop this programme which, in a certain sense, culminates the lines of convergence pointed out by Husserl's levels and which Steinbock has reconstructed with precision. For my part, I believe that this sound starting point can only be advanced through what I call a situational phenomenology. Because knowing the moment we were never in is only feasible if we construct the constellation of loss in such a way that it is phenomenologically relevant or evident to us—those who construct it. This implies the condition that the constructed situation is still existentially binding for us. Only then are the promises of concreteness that generative analysis promises fulfilled. I wish to refine such an introduction to this situational phenomenological anthropology in my next works, and to do so in the spirit of Blumenberg's monumental philosophical legacy.

References

Blumenberg, H. (2006). *Beschreibung des Menschen*. Suhrkamp.

Blumenberg, H. (2010). *Theorie der Lebenswelt*. Suhrkamp.

Bruzina, R. (2004). *Edmund Husserl and Eugen Fink*. In *Beginnings and Ends in Phenomenology, 1928–1938*. Yale University Press.

Cassirer, E. (1925). *Sprache und Mythos*. B. G. Teubner.

Cassirer, E. (2020). *The Philosophy of Symbolic Forms*. Routledge.

Fischer, J. (2008). *Philosophische Anthropologie. Eine Denkrichtung des 20. Jahrhunderts*. Alber.

Gault, S. (2017). Tracing Merleau-Ponty's Passage to Ontology. *Chiasmi International, 19*, 345–369.

Heidegger, M. (1983). *Die Grundbegriffe der Metaphysik: Welt, Endlichkeit, Einsamkeit*. V. Klostermann.

Husserl, E. (1950a). *Cartesianische Meditatationen und Pariser Vorträger*. Martinus Nijhoff. [CM, Hua 1].

Husserl, E. (1950b). *Die Idee der Phänomenologie. Fünf Vorlesungen*. Martinus Nijhoff. [Hua 2].

Husserl, E. (1950c). *Ideen zu einer reinen Phänomenologie und phänomenologischen Philosophie. Erste Buch: Allgemeine Einführung in die reine Phänomenologie*. Martinus Nijhoff. [Hua 3].

Husserl, E. (1952). *Ideen zu einer reinen Phänomenologie und phänomenologischen Philosophie. Zweite Buch: Phänomenologische Untersuchungen zur Konstitution*. Martinus Nijhoff. [Hua 4].

Husserl, E. (1954). *Die Krisis der europäischen Wissenschaften und die transzendentale Phänomenologie. Eine Einleitung in die phänomenologische Philosophie*. Martinus Nijhoff. [Krisis, Hua 6].

Husserl, E. (1956). *Erste Philosophy (1923/1924). Erster Teil: Kritische Ideengeschichte*. Martinus Nijhoff. [Hua 7].

Husserl, E. (1959). *Erste Philosophy (1923/1924). Zweitee Teil: Theorie der phänomenologischen Reduktion*. Martinus Nijhoff. [Hua 8].

Husserl, E. (1973a). *Zur Phänomenologie der intersubjektivität. Texte aus dem Nachllass. Erster Teil: 1905–1920*. Martinus Nijhoff. [Hua 13].

Husserl, E. (1973b). *Zur Phänomenologie der intersubjektivität. Texte aus dem Nachllass. Dritter Teil: 1929–1935*. Martinus Nijhoff. [Hua 15].

Husserl, E. (1987 [1911–1921]). *Aufsätze und Vorträge*. Kluwer. [Hua 25].

Husserl, E. (1999). *Hua-Dok III*. Springer.

Husserl, E. (2005). *Späte Texte über Zeitkonstitution (1929–1934): Die C-Manuskripte (Husserliana: Edmund Husserl – Materialien)*. Springer.
Husserl, E. (2006). Die kopernikanische Wende der kopernikanischen Wende. Die ursprüngliche Arche Erde. In J. Dönne & E. Günzel (Eds.), *Raumtheorie. Grundlagentexte aus Philosophie und Kulturwissenschaften* (pp. 153–165). Suhrkamp.
Husserl, E. (2008 [1961–1937]). *Die Lebenswelt. Auslegungen der Vorgegebenen Welt und ihrer Konstitution, Texte aus dem Nachlass*. Springer. [Hua 39].
Husserl, E. (2014 [1908–1937]). *Grenzprobleme der Phänomenologie: Analysen des Unbewusstseins und der Instinkte. Metaphysik. Späte Ethik, Texte aus dem Nachlass*. Springer. [Hua 42].
Karädi, È., & Feket, È. (Ed.) (1982). *Georg Lukàcs. Briefwechsel 1902–1917*. J. B. Metzler.
Lévy-Bruhl, L. (1929). *Die Geistige Welt der Primitiven*. Brückmann.
Lukács, G. (1974). *Heidelberger Philosophie der Kunst (1912–1914)*. Luchterhand.
Lukács, G. (1975). *Heidelberger Ästhetik (1916–1918)*. Luchterhand.
Merleau-Ponty, M. (1968). Husserl aux limites de la phenomenology. In *Résumés de cours, Collège de France, 1952–1960* (pp. 159–171). Gallimard.
Merleau-Ponty, M. (1998). *Notes de cours sur* "L'Origine de la géométrie". Presses Universitaires de France.
Plessner, H. (1931). *Macht und menschliche Natur, Ein Versuch zur Anthropologie der geschichtlichen Weltansicht*. Junker und Dünnhaupt Verlag.
Plessner, H. (1975). *Die Stufen des organischen, Einleitung in die Philosophische Anthropologie*. Walter de Gruyter.
Plessner, H. (2002). *Grenzen der Gemeinschaft. Eine Kritik des sozialen Radikalismus. Cover: Grenzen der Gemeinschaft*. Suhrkamp.
Sowa, R. (2008). Einleitung Herausgebers. In E. Husserl, (Ed.), *Die Lebenswelt. Auslegungen der Vorgegebenen Welt und ihrer Konstitution, Texte aus dem Nachlass* (pp. XXV–LXXXI). Springer.
Steinbock, A. J. (1995). *Home and Beyond, Generative Phenomenology after Husserl*. Northwestern University Press.
Steinbock, A. J. (2022). *Mundo familiar y mundo ajeno. Fenomenología generativa*. Sígueme.
Villacañas Berlanga, J. L. (2021). Los dioses del instante y el Ereignis. In R. Navarrete Alonso & C. Basili (Eds.), *Arcana del pensamiento del siglo XX* (pp. 17–39). Herder.
Villacañas Berlanga, J. L. (2024). *Ortega y Gasset, una experiencia filosófica española*. Guillermo Escolar y Mayo.
von Wolzogen, C. (n.d.). Lebenswelt 1919 oder Die Wissenschaft von den Trivialitäten-Heidegger liest Husserl. Eine Mizelle. *Akademie.edu*. https://www.academia.edu/41405145/Lebenswelt_1919_oder_Die_Wissenschaft_von_den_Trivialit%C3%A4ten_Heidegger_liest_Husserl_Eine_Miszelle

José Luis Villacañas Berlanga is a professor of Philosophy at the Complutense University of Madrid, where he has been director of the Department of Philosophy and Society from 2009 to 2022. Previously he was a professor of Moral and Political Philosophy at the University of Murcia. Doctor in Philosophy from the University of Valencia, with a thesis on *Empirical realism and transcendental idealism in Kant: the levels of its use and justification*, he has specialized in German philosophy from the Enlightenment to the present day of Blumenberg's thought, passing by the work of Weber and Schmitt. At the same time, he discusses the relationships between Spanish and European thought, analyzing the influence of Kant, Weber or Carl Schmitt in Spain. He directs the magazines *Res Publica, Magazine of the History of Political Ideas*, and *Annals of the History of Philosophy Seminar*. His latest books are *Luis Vives* (2021); *Ortega y Gasset. Una experiencia filosófica española* (2023); *Sublime psíquico y educacion estética* (2023); *Justicia viva. El pensamiento de Giorgio Agamben* (2024), and *Max Weber en contexto. Filosofía y ciencia social tras las sendas de Kant* (2024).

Defense of 'Expanded Reason' in Hans Blumenberg: The Case for the Science of Boredom

Josefa Ros Velasco

Abstract The German philosopher Hans Blumenberg always advocated the need to adopt an open and integrating view of the wisdom from different disciplines of knowledge when delving into the intricacies of the reality in which we find ourselves immersed, not only as a formula to try and embrace it in theoretical terms, but as a necessary resource to live in a meaningful way. Particularly when speaking of the human being, Blumenberg warned that the aspects that make up our life experience can only be apprehended based on a methodological approach in which there is room not only for scientific rationality, but also for humanistic and even spiritual contributions. This vision coincides with the commitment to what has recently been described, based on the various calls for multidisciplinary research made by Pope Benedict XVI, under the term 'expanded reason.' A review of Blumenberg's methodology in the pursuit of knowledge of things can help us to better understand what is at stake in the demand that this term has been making since the mid-twentieth century and, more than ever, in the present era. In this chapter, I will explore what it means to approach the human phenomena that take place in the *reality in which we live* from the perspective of expanded reason. However, I will also observe how the philosopher has put into practice this expanded reason approach to the study of an aspect of human life to which he attributed unparalleled importance in the history of humanity: the experience of boredom.

This chapter has been written within the framework of the European Union's Horizon 2020 research and innovation programme under the Marie Skłodowska-Curie grant agreement No 847635.

J. Ros Velasco (✉)
Department of Humanities, ESIC University, Madrid, Spain
e-mail: josros@ucm.es

A. Fragio et al. (eds.), *Hans Blumenberg's History and Philosophy of Science*, Studies in History and Philosophy of Science 63,
https://doi.org/10.1007/978-3-031-78251-0_6

1 Introduction

On Thursday, 19 October 2006b, at Verona Exhibition Centre, Pope Benedict XVI gave the speech known as *Address of His Holiness Benedict XVI to the Participants of the Fourth National Ecclesial Convention*, in which he warned of the urgency of establishing a dialogue among the sciences, philosophy, and theology to bring to fruition the endless exercise of knowing the reality in which we find ourselves immersed and, specifically, to try an approach to the human phenomenon, "in full respect for the methods proper to them and of their reciprocal autonomy, but also in the awareness of the intrinsic unity that holds them together."

Inspired by the words of the now-former Pope, Francisco de Vitoria University coined the term 'expanded reason,' which also gave its name to the Expanded Reason Institute of Spain, to refer to "that way of conceiving and using reason that seeks to know reality in all its breadth, through dialogue between the sciences, philosophy and theology" (2024, a. trans.). In accordance with the mission that is made explicit on the Institute's website (2024), in which Ratzinger's words are also cited, this time following the *Lecture of the Holy Father: Faith, Reason and the University. Memories and Reflections*, September 12, 2006a, in Regensburg:

> In the face of reductionism of any kind (empiricism, utilitarianism, fideism...), this vision advocates opening reason to all aspects of reality, to all forms of knowledge, and to all dimensions of the person, which requires making room for the possibility of the "encounter of faith and reason, between genuine enlightenment and religion" *[sic]*. It thus requires bringing into play the whole person: reason and faith, science and conscience, intelligence and heart. In the face of the fragmentation of knowledge and hyper-specialisation, it invites us to seek a harmonious synthesis of knowledge (a. trans.).

The German philosopher Hans Blumenberg always advocated the need to adopt an open and integrating view of the wisdom from different disciplines of knowledge when delving into the intricacies of the reality in which we find ourselves immersed, not only as a formula to try and embrace it in theoretical terms, but as a necessary resource to live in a meaningful way. Particularly when speaking of the human being, Blumenberg warned that the aspects that make up our life experience can only be apprehended based on a methodological approach in which there is room not only for scientific rationality, but also for humanistic and even spiritual contributions. Although Blumenberg never used the term 'expanded reason' as such, a review of his methodology in the pursuit of knowledge of things can help us better understand the demand that the term has been making since the mid-twentieth century and, more than ever, in the present era (Ros Velasco, 2017a).

In this chapter, I will explore what the expanded reason approach to the human phenomena that take place in the *reality in which we live* consists of. We will also observe how the philosopher has put into practice this expanded reason approach to the study of an aspect of human life to which he attributed unparalleled importance in the history of humanity: the experience of boredom (Ros Velasco, 2017b, 2019). Boredom is an everyday phenomenon that torments us when reality does not meet our expectations. We are very familiar with it. It visits us all at some time, in its simplest and most fleeting form or its deep and lasting expression. We even feel it as a group. The pain it causes us predisposes us to do anything to end it. From its suffering emanates human creativity and also the worst monsters. Its experience is pathological to the point of being attributed to illness. However, boredom is only a symptom. Its mission is to alert us to the fact that our relationship with our environment is damaged. Our responsibility is to delve into the root of the problem and repair it.

The issue of boredom became the subject of scientific and academic research when, at the beginning of the last century, the mental health sciences began to pay attention to it in the context of occupational psychology. Since then, psychologists and psychiatrists have turned their attention to boredom in an attempt to scientifically demonstrate what it consists of, what its main causes and consequences are, and how we can rid ourselves of the discomfort it causes us, avoiding unhealthy responses. Many of the assumptions from which boredom is approached within this scientific framework are to be found, however, in the philosophy and theology of previous centuries, from Antiquity to the beginning of Modernity, inevitably passing through the Middle Ages (Ros Velasco, 2022, 2024a, b).

The efforts that Western philosophers and theologians, for example, have made throughout history to clarify the experience of boredom from an anthropological, epistemological, ethical and spiritual point of view are rarely acknowledged. The reflection in this chapter represents a reconciliation between scientific and humanistic rationality through the exposition of the philosophical, theological, literary, anthropological, sociological, and medical narratives that Blumenberg coordinates to study boredom and which highlight the interconnection between all of them and the need for an interdisciplinary approach when dealing with a multifactorial phenomenon such as the one we are dealing with. Ultimately, the aim is to show that certain relevant aspects of human life, such as the experience of boredom in this case, can only be understood through an approach based on expanded reason that combines scientific rationality with philosophical and theological dialogue; an approach that was the hallmark of the Hanseatic philosopher's way of practicing science.

2 The Limits of Scientific Rationality and the Need for Expanded Reason

One of the main problems facing any science that tries to explain human phenomena is that it cannot always satisfactorily respond to all the epistemic demands placed on it without other discourses. Constant advances and paradigm shifts are permanently shedding light on the human scenario, making it impossible to establish a last word or an unquestionable truth regarding some of the aspects we are trying to learn more about. Therefore, the approach to the human can never be definitive; it is always under construction, which makes it difficult to approach from the unique perspective of a single discipline (Ros Velasco, 2017b).

However, nowadays, we place a great deal of trust in scientific rationality as the bearer of irrefutable truth in its exercise of unravelling the mysteries of human nature. While it is true that many of these conditions can be reproduced in laboratories to test the hypotheses that are drawn up, some of the key points for apprehending them resist being approached solely through scientific rationality, we must, therefore, approach them using hypotheses and speculations with the means made available to us by other forms of knowledge, such as those from the humanities (Ros Velasco, 2017b).

Despite everything, these realities need to be understood, as it is through this understanding that we can achieve a certain horizon of meaning from which to orientate our praxis (Ros Velasco, 2017b). It is not a coincidence that the most difficult questions to explain are precisely those we most need to understand, as Kant pointed out in the introduction to the prologue to the first edition of the *Kritik der reinen Vernunft* (KrV A vii):

> Human reason has the peculiar fate in one species of its cognitions that it is burdened with questions which it cannot dismiss, since they are given to it as problems by the nature of reason itself, but which it also cannot answer, since they transcend every capacity of human reason (1998, p. 99).

Those that refer to realities that are difficult to pronounce on, and seem to escape human understanding make us feel most powerless, cause us uncertainty and push us to desperately seek justification, "it is a question of the necessary understanding of ourselves" (Blumenberg, 2013, p. 248, a. trans.). These kinds of questions denote what the philosopher Hans Blumenberg (1985) called the *absolutism of reality* (*Absolutismus der Wirklichkeit*), one from which we must gain distance to live a meaningful life and against which we erect discourses that go beyond scientific rationality to generate horizons of meaning (Ros Velasco, 2012).

According to the Blumenbergian hypothesis of the absolutism of reality, human beings are immersed in the most despotic scenario, being part of that omnipotent nature, of the unlimited and arbitrary existence of the universe, which surrounds us and escapes our cognitive ability. This reality that is unknown to us in its original form, that resists being pigeon-holed among the objectivist pretensions of the human gaze, is the amorphous and disfigured whole that appears unfathomable to us, that causes us impotence after the attempt at delimitation, revealing our difficulty in defining what is indefinable and giving rise to constant human reflection that concludes in such an unfortunate whereabouts, in fear and anguish of knowing that the world exists beyond us, indifferent to our own existence (Blumenberg, 1985, 1986, 2013; Ros Velasco, 2012, 2017b).

What we call absolutism of reality, which Blumenberg defines in *Arbeit am Mythos* (1985, p. 4) as "the totality of what goes with this situational leap, which is inconceivable without super-accomplishment in consequence of a sudden lack of adaptation," puts us under pressure and torments us, making us feel disconcerted and uncomfortable in our dealings with the world, disoriented by a relationship dominated by imbalance and inadequacy in which we cannot outline more truth than that coming from human creation. This is the situation that characterizes our relationship with reality, given the contingent nature that dominates the gaze towards what is absent behind the horizon, which invites us, ultimately, to give free rein to a world of signs and images we ourselves create to acquire existential meaning (Cassirer, 1965), to redirect the dissonances with a world that at all times presents itself to us in an arrogant and threatening way (see also Ros Velasco, 2012, 2017b).

In order to respond to certain current challenges that cause us disorientation and endanger our well-being, we elaborate non-scientific narratives *sensu stricto* intending to reach an agreement with the absolutism of reality. It is precisely the mediation that is established through the symbolic edifice between man and reality that makes it possible for the world to represent and signify a meaningful scenario for our species, and for us to sketch some form of knowledge regarding it, correcting the existing discordances and preventing them from growing out of all proportion until they lead us to meaninglessness and chaos. In other words, to cope with the malaise caused by the absolute reality, we aim to gain distance from it as much as possible, doing everything to turn it into something more familiar and accessible, by either giving it a name, explaining it through a story or making use of whatever rhetorical stratagems are necessary. Through as many tricks as necessary we try to reach "the substitution of the familiar for the unfamiliar, of explanations for the inexplicable, of names for the unnameable" (Blumenberg, 1985, p. 5). We have recourse to symbolic forms, a system to suppress arbitrariness, to transform the parts of reality we find unfamiliar and menacing into accessible, familiar phenomena, identifying the unknown through its designation (Blumenberg, 1985; Ros Velasco, 2012, 2017b).

For example, metaphors replace the unavailable with the available, as part of the deconstruction process of the absolutism of reality that has already begun to make our home threatening. In his case, the absolute metaphors, which Blumenberg speaks about in *Paradigmen zu einer Metaphorologie* (2010), refer to representations that man creates of his existence and of the world whose function is to facilitate the understanding of the most inaccessible aspects of reality. Blumenberg has tried to justify and vindicate the role of metaphors as instruments through which the human intellect sets in motion the process of ordering sensory chaos, considering them not as mere preliminary stages in the formative process of concepts, but as symbolic units whose capacity to generate meaning reaches even where the latter are incapable of doing so. They are essential and unavoidable elements in man's relationship with reality; they are an irreducible form of thought, even for the scientific and philosophical disciplines, given that they possess their own significance in our way of relating to things without which we could not live with a certain ontological security (Ros Velasco, 2012, 2017b).

This is how human beings manage to dominate arrogant, unfathomable reality, especially when our intellect needs to tackle objects that require too much of it. Through rhetoric we produce what is necessary to reach a consensus with the absolutism of reality, to distance it and thus reduce our discomfort in the face of the unknown, or, in other words, to bring us closer to reality by making it favourable and more comprehensible. Rhetoric and its forms, such as metaphors or narratives, fulfil a regulating task, since they allow us to draw lines of connection between the dominating nature and our insignificant perspective, preventing us from falling into terror, nullity and disorientation in the face of the mysteries of the unknown. Rhetorical figures allow humans to reach a consensus on what is true in reality, according to what is useful for their survival, and try to ensure that there is no contradiction or breakdown in what is accepted, so that it becomes contagious, generating a base of common convictions, "that unknown which appears on the periphery is constantly buffered by actions aimed at dominating it: giving it a name, integrating it in metaphorical form, finally subordinating it and classifying it conceptually" (Blumenberg, 2013, p. 147, a. trans.).

These narratives fulfil a regulatory function, since they allow us to draw lines between the dominating nature and our insignificant perspective and reach a consensus on what is truth, based on what is useful for our survival, avoiding the descent into fear, nullity and disorientation in the face of the mysteries of the unknown (Blumenberg, 1985; Ros Velasco, 2012, 2017b). The value of these non-scientific explanations or narratives is measured by the extent to which they succeed in generating meaning. It is not a means by which we are satisfied; it is not a substitute for scientific truth; it is all that can be expected of human beings on certain occasions. We are thus confronted with a type of unverifiable knowledge that has been consistently rejected by those who trust in the epistemic ideal of scientific rationality. However, the strength of hypotheses and non-scientific constructs lies in the fact that they answer questions of great interest to humanity and, we cannot do without what they offer us: an integral conception of reality. The status that corresponds to any non-scientific hypothesis or narrative is that of verisimilitude (Blumenberg, 1985), which does not represent the truth, but neither is it a lie, and which serves to satisfy our needs for knowledge and meaning. It is, in Blumenberg's own words, something like a "calculated admission of the unlawful, whose license is due to the fact that otherwise there would be nothing at all" (2013, p. 12, a. trans.). Following the principle of insufficient reason, such approaches are justified by their usefulness in praxis and require a certain "willingness to let certain degrees of indeterminacy pass" and a willingness to accept "the rationality of the absence of substantiation" (Blumenberg, 2013, pp. 98, 13, a. trans.; see also Ros Velasco, 2017a).

However, to some degree, for these approaches that cannot be framed within scientific rationality to fulfil their task of shedding light and generating understanding, they must not contradict those aspects that science has been able to demonstrate following its method (Ros Velasco, 2017b). Blumenberg (2010) admits that we cannot retreat to fiction in those cases where formulas are possible. We do not have to resort to metaphorical figures or myths to answer questions that can be answered more precisely and rigorously. That is why, as Barbara Merker warned, the narrative must be informed by and consistent with scientific findings (2013, p. 125):

> Fictional narrative (like Alsberg and Blumenberg's primal scenes), such as the history of literary fiction (like Conrad and Wertheimer's novel) or discursive speculation (like Schrenk's extrapolations) must not be arbitrary. They must include current knowledge and not infringe or violate it. The decisive factor is the plausibility of the subjunctive sentence: what could have been, even if it has never been proved (a. trans.).

In short, and now recovering Kant's words, in *Muthmaßlicher Anfang der Menschengeschichte*, about how the science of history is constructed (AA 8:109):

> It is surely permissible to *insert* here and there conjectures into the *progression* of a historical account, in order to fill the gaps in the records. For what precedes the gaps (the remote-cause) and what follows them (the effect) give a fairly reliable clue to the discovery of the intermediate causes, which are to make the transition intelligible. But to originate a historical account *from* conjectures alone would seem to be not much better than to draft a novel. Indeed this could not be called a *conjectural history* but rather a mere piece of fiction (2007, p. 163).

Blumenberg (1985, 2010) never had any objection to this approach; the problems always came from scientists' denial of the knowledge provided by narratives they considered of lesser value. As he makes clear in countless examples, science and fiction have always met and made use of each other. One of them is the example in which he states that thanks to the level of realism that the German scientist Kepler had given to geometrical metaphors, he ended up obtaining the laws from the representative scheme of the sphere as a projection of the centre. There was, therefore, a passage from the representative scheme to the conceptual conception of gravitation. Through this case, in which we find Kepler endowing the Sun with a whole series of metaphysical attributes through which he can approach the ideas of Divine creation, Blumenberg shows how metaphor allows the conquest of new possibilities and means of rational construction that are decisive for the subsequent theoretical process. Another example can be found in the case of the German philosopher Friedrich Christoph Oetinger, who, by exchanging the traditional roles of Nature and God, succeeded in deducing the approximately circular orbits of celestial bodies whose motion had to be rectilinear thanks to the intervention of God's will. In the same way, the metaphor of the polygon of forces gave birth to the theory of the state division of powers in the political thinker Montesquieu, embodied in his work *De l'esprit des lois*, based on the situation in which autonomous factors act in their own interests and at the same time try to preserve the interest of the whole in which they are immersed (see also Ros Velasco 2017a, b).

Although narratives of a philosophical or theological nature can frequently be left aside because the reality they explain finds a justification in scientific rationality, there are, however, questions that can never be answered in these terms alone (Blumenberg, 2010). The question of the experience of human boredom, as will be seen in the following sections, is one of them (Ros Velasco, 2017b). Blumenberg claims the role of the philosophical and theological approach as a tool by which the human intellect sets in motion the process of generating meaning where scientific rationality alone is insufficient. This is in line with the presupposition of expanded reason according to which non-scientific rationality must rely on and be coherent with scientific discoveries, and must also continue to draw on the knowledge offered by science so that, in a joint effort, the synthesis of knowledge can provide answers to the anthropological, epistemological, ethical and spiritual questions underlying certain phenomena of human experience, as in the case of boredom.

Scientific rationality has traditionally rejected twinning with the humanistic disciplines, presuming to have achieved an epistemic status, a degree of objectivity, which the latter will never be able to attain, forgetting that the concepts with which it works cannot do without the former when it comes to answering certain questions. Despite their manifest indebtedness, the reconciliation between the parties has constantly failed due to the difference in epistemological expectations: science promises eternal, demonstrable truths, while the humanities are committed to a temporary consensus that allows for a multiplicity of voices. However, we seem to have forgotten the philosophical and spiritual component on which science rests, which needs to answer the questions most deeply rooted in the human spirit (Ros Velasco, 2017a).

As participants in a scientifically advanced society, we admit, like Blumenberg, that scientific rationality has brought much to our lives: the capacity for prediction, increased control and a certain ontological security. The sciences have built their achievements on the assumption of attaining truth through empirical observation, trying to dispense with any form of symbolic human construct. However, both the truth and the objectivity that we achieve through science are still the fruit of our own ordering, of the formative function of our intellect that determines reality in an absolute way. Scientific rationality is also, in the end, the result of the work that our understanding does in its effort to apprehend the plurality that is given to us, one that cannot only be carried out through logic, but also through our imaginative faculty.

We cannot expect science to satisfy our deep-rooted claims and expectations of meaning on its own. When it dispenses with philosophical and theological discourse, in contrast, what happens is that it makes the absolutism of reality more evident. Science has objectified the world and is now mute in the face of questions concerning the world itself and questions concerning nature and human life. In solitude, scientific rationality cannot prevent us from feeling the emptiness of the cosmos. Scientific knowledge forces man to be aware of a situation that is not easy to bear. In this regard, it is doubtful whether science can serve, as philosophy or theology do, to guide our praxis (Ros Velasco, 2017a). What is important, after all, is not so much to present true and unquestionable knowledge, but that it is meaningful (Ros Velasco, 2017b). Thus, it may be all the more irrational to insist on proceeding solely on the assumptions of scientific rationality, simply for the sake of reaching definitive truths, than to accept the approach advocated by expanded reason capable of coexisting with the plausible.

The analytical proposal that seeks to do away with all kinds of statements for which no evidence can be found and which, therefore, cannot be verifiable, would entail the overthrow of the system on which human praxis is based (Ros Velasco, 2017a). For this reason, Blumenberg (1985, 2010) advocated revaluing the strength that lies in the instruments with which the humanities approach the reality in which we find ourselves immersed, embracing the approach that defends expanded reason with which we try to broaden the horizon of scientific rationality in a dialogue with philosophy and theology that conveys to us the truths that nestle in the hearts of human beings and that aspires to give meaning to human life in society. As shown below, the question of the experience of boredom is precisely at this crossroads, to which Blumenberg (see Ros Velasco, 2019, 2025a) sought to find a solution by committing to the synthesis of knowledge that arises from the proposal of expanded reason.

3 Boredom from Scientific Rationality and Expanded Reason

Recently, a space has opened up in our public and academic dialogue for debate regarding the common phenomenon that human beings get bored. Yes, we get bored. We get bored by situations that we find insufficiently stimulating, we get bored by repeated exposure to the familiar, we get bored when we don't know what to do—even though we know we want to do something—and we get bored when we do things that are not meaningful to us. We get bored as individuals, but we also get bored as a community. We are all bored, without exception, even those who boast that they are never bored, boasting of their ability to delight in every wonder of creation, or those who dare not admit their boredom for fear of being labelled as silly, uncurious or unproductive slackers.

We all get bored occasionally when we get involved in situations or activities that are not very stimulating, lacking in interest, repetitive or forced, insignificant, in the end, that do not respond to our needs for cognitive excitement. It is a state of discomfort that we experience more or less frequently, in all places and at all times, depending both on exogenous factors, which obey the possibilities of the environment, and endogenous factors, related to our own individual personality. Boredom is thus a multifactorial phenomenon that must be approached from a broad multidisciplinary perspective, as has been done throughout history from philosophical, theological, literary, and also medical narratives (Ros Velasco, 2022, 2024a).

Undoubtedly, understanding what boredom consists of, what its main causes and consequences are, and learning how to manage it optimally, avoiding maladaptive responses to its suffering, is of utmost importance not only to achieve a robust knowledge regarding a state that has characterised us since time immemorial, but also to guide our praxis. It is an exercise we have been trying to conquer for centuries because it helps us understand our way of experiencing reality at an individual and collective level, but also because of the need to avoid the negative correlate that comes from constantly suffering from boredom (Ros Velasco, 2022, 2024a, b).

To date, the literature has shown that boredom, in its most pathological form, is sometimes associated with several mood disorders such as rage, anger, disaffection, apathy, depression, anxiety, stress or alexithymia; with behavioral disorders such as suicide, delinquency and criminality, rebelliousness and provocation, bestiality, impulsivity, reckless driving or addictions to drugs, sex, gambling, the Internet, mobile phones or eating and sleeping disorders; with personality disorders such as hysteria and narcissism; or with pathological states of self-consciousness, identity and introspection; and finally with mental illnesses such as psychosis, schizophrenia, paranoia, Alzheimer's, Asperger's syndrome and bipolar disorder, among many, many others (Ros Velasco, 2022, 2024a, b).

Over the last century, the issue of boredom was explored almost exclusively from the paradigm of scientific rationality, with the mental health sciences of psychology and psychiatry playing a predominant role. The pathological and painful nature of boredom has also meant that numerous thinkers interested in mental health have also looked to this state. It was in the early twentieth century when medicine took over from the humanities regarding the study of boredom, almost completely erasing what had come before. From humanities, boredom was understood as an illness, but only in metaphorical terms. In reality, it was no more than the consequence of the decadence of the West. However, psychologists and psychiatrists soon placed the focus on the subject who is bored, forgetting the social context (Ros Velasco, 2022, 2024a, b).

When the first psychological schools of this period came onto the scene, from which boredom was understood through its relationship with psychic disorders, a phenomenon of breakdown began to occur in the synthesis of knowledge that, until then, the different disciplines involved in its study had reached in order to achieve a holistic approach to its nature and to guide our praxis. Psychologists and psychiatrists of the twentieth century began to work on a definition of boredom from which to diagnose pathologies related to this state and offer appropriate treatments that completely disregarded other branches of knowledge that addressed the most painful experience of boredom in its entirety, taking into account issues related to anthropology, epistemology, ethics or metaphysics (Ros Velasco, 2022, 2024a, b).

Examples of this were the psychological studies of Theodor Lipps, who described the psychodynamic theory according to which boredom was a pain provoked by "the conflict between the internal need for an intense mental activity and the lack of stimulation or the incapacity of the subject to stimulate it" (1909, pp. 337–338, a. trans.). From his point of view, the lack of stimulation did not have to correspond to the lack of stimuli from an external source: the subject could be *incapable* of feeling them due to some pathological feature related to their personality. He talked about individuals with a "weak will" (*Willensschwachen*), of "despondent" (*Mutlosen*) (Lipps, 1909, p. 338, a. trans.) individuals who were unable to be interested in a goal, in pointing to it and in using the necessary energy to achieve the desired mental effect. They lacked the power of self-persuasion to be attracted by things, perhaps due to a problem of "lack of concentration" (Lipps, 1909, p. 338, a. trans.). Others like the psychologist Hugo Münsterberg (1913) carried out the first psychological studies on boredom in factories. He said that the monotony which gave rise to boredom depended not so much on repetitive work as on the individual's *disposition*. In view of the situation, psychiatrist Joseph Barmack tried to find a form of mitigating boredom in the workplace by intervening directly on the psychological state of the bored person using amphetamines and other stimulants (1939a, b). The Austrian psychiatrist Edmund Bergler (1945) also asserted that boredom was as a disease entity on a level with erythrophobia and depersonalization, even leading to suicide.

During the 1950s, many specialists focused on delimiting the nature and causes of endogenous pathological boredom (Bieber, 1951; Greenson, 1951, 1953; Stuart, 1951), even trying to clarify its psychobiological substrate (Levinger, 1951). From then on, the predominant tendency was to analyze the reaction to chronic boredom and its possible anomalies. Specialists began to explain the reasons for the most problematic forms of boredom by measuring physiological changes in subjects exposed to unstimulating situations (Berlyne, 1960, 1967; Cooper, 1968).

In the 1980s, it was the cognitive theories focusing on the individual's perception of the degree of stimulation of the environment that were most prevalent. For example, O'Hanlon (1981) noted that boredom came from a certain disorder/alteration of the normal functions of the brain, taking place in the temporal and spatial confines of a monotonous task. For the first time, boredom studied from the perspective of its cognitive components began to be analyzed according to a demographic factor as relevant as age. A study carried out by Sundberg and Bisno (1983) was the starting point for this trend, which would be bolstered in the following years by the works on boredom among children and adolescents. Boredom was also observed in relation to stress, fear, impulsivity, and violence, among other psychotic factors (Sommer et al., 2021). In addition, the first studies appeared which blamed boredom for excessive consumption of alcohol and other drugs (Ros Velasco, 2022, 2024a, b).

From a psychoanalytical approach, what began to take precedence during the 1980s was the demand to find methods of diagnosing the disorders related to boredom (Maggini & Dalle Luche, 1987; Savitz & Friedman, 1981). This phenomenon fostered the emergence of many articles on therapeutic treatments to ease boredom, such as those by Revers (1983) or Hozier (1988), which postulated a type of therapy to reconnect with one's own emotions which had been lost over life. Others, like Patrick (1982), moved to recreational therapy, which would take on an essential value in treatments against boredom in the following decade. At this time there also appeared the first scales of measurement. Before the 1980s, there was already a subscale for measuring boredom, the Boredom Susceptibility Scale (BSS) of 1964, as part of the Sensation Seeking Scale (SSS) of Marvin Zuckerman et al. (1964). However, Farmer and Sundberg (1986) marked a turning point in this field by designing and publishing a new scale to assess the proneness to boredom, not only concerning external stimuli but also the capacity for internal self-stimulation. This was the Boredom Proneness Scale (BPS) of 1986. This scale was applied in countless studies led by a research group at the University of West Florida, headed up by Watt and Vodanovich. In these they showed, among other things, that subjects prone to boredom saw their perception of time and their ability to generate self-stimulation altered, and that the propensity to boredom was the cause of increased aggressiveness (1992a, b, 1999).

The West Florida group occupied and continue to occupy a core place in the studies of boredom and represent a reference point, but many others across the world paid attention to boredom from mental health disciplines in the 1990s, and since the beginning of the new millennium, with contributions on the relationship between boredom and mood, behavior and personality disorders, problems in the workplace, demographic traits such as age, and, finally, mental illnesses (Ros Velasco, 2017b,

2022, 2024a). Once the twenty-first century arrived, some laboratories opened their doors, such as the Boredom Lab, the Danckert Lab, and the Socoul Lab in which, beyond theoretical analysis, some colleagues are dedicated to using fMRI techniques to understand the roots of boredom (Danckert et al., 2018).

The most surprising is that, at the end of the twentieth century and the beginning of the twenty-first, a total confusion and lack of agreement was reported with respect to a common state that had become a chimera for mental health sciences (Ros Velasco, 2017c, 2022, 2024a, b). The competition between the different approaches to clarify what boredom was made it impossible to compare and harmonize the evaluation methods, thus reducing the possibilities of agreement between the parties and preventing a real comprehension of its nature (Vogel-Walcutt et al., 2012). Despite the abundance of definitions of boredom since the beginning of the century by mental health experts—in works such as the *Dictionary of Psychology* by Warren (1934), the *Dictionary of Psychiatry* by Thakurdas and Thakurdas (1979), the *Corsini's Encyclopedia of Psychology* (1999), the *Psychiatric Dictionary* by Campbell (1996), or the *APA Dictionary of Psychology* (VandenBos, 2007)—boredom continued to be "an ill-defined scientific concept" (O'Hanlon, 1981, p. 53).

Researchers from the mental health sciences complained that the lack of knowledge regarding boredom stemmed from the lack of interest in this phenomenon and the paucity of existing literature (Ros Velasco, 2017c). However, any researcher from the humanities could see that this was untrue. Furthermore, the problem that had been incurred in recent times in Boredom Studies, conducted from the mental health sciences, was that the long tradition existing since Antiquity, which had analyzed this phenomenon through the synthesis of different knowledge hailing not only from philosophy and theology, but also from medicine, had been ignored (Ros Velasco, 2024b). One that would have been able to offer a holistic answer to the nature of boredom—including not only the question of physiology, but also anthropology, epistemology, ethics and metaphysics—if it had continued to use the perspective of expanded reason that had hitherto been put into practice—and from which Blumenberg undoubtedly approached the question of boredom.

There is a clear relationship between the principles that philosophers, theologians, literati and physiologists of all ages have established about the experience of boredom and those of today's psychologists and psychiatrists who went their separate ways. Without doubt, the disciplines of mental health have driven forward the studies on boredom in a very short time, from a perspective complementing that offered by the humanities and sociology. Examining in detail the concept of endogenic and individual chronic boredom, whose precedents are dispersed across the history of Western thought, has constructed a new basis for this branch of knowledge. However, we cannot neglect the evidence that there has been a clear transfer between disciplines over the centuries, which indicates that multidisciplinary work may be key for an understanding of a phenomenon that is as many-sided as that of boredom, particularly with a view to avoiding the negative consequences that may result from suffering from it (Ros Velasco, 2024b). We cannot ignore the contributions that philosophical and theological discourses have made when it comes to understanding boredom, from those made by classical authors like Horace,

Lucretius, and Seneca, to contemporary philosophers such as the American Andreas Elpidorou. The following brief tour serves as an example (Ros Velasco, 2023).

During the Roman empire, these mentioned philosophers wrote about the experience of boredom in its simplest form (*fastidium, satietas, desidia*) and in its most complex state (*taedium vitae*), defining it as heaviness, lentitude or satiety, due to an excess of idleness, leading to mental disorders, irritation and apathy (Seneca, 2004, 2009, 2014; Lucretius, 2008; Horace, 2011). In the Middle Ages, boredom, known as *acedia*, meant the alienation of the soul for St. Augustine of Hippo (2006, 2007, 2008) and Ponticus (2006) described it as despondency and sadness compared with a fulfilled life; Cassian (1999, 2015) used the terms weariness, disgust and anxiety of the heart to characterize it; John Chrysostom (PG 47.425-426) indicated that it was a destructive spiritual condition generated by a lack of perceived sense and desperation; St. Jerome (PL 22.1081-82) assimilated it with melancholy caused by loneliness; Hugh of Saint Victor (PL 176.525-526) mentioned that it was a disturbance of the mind caused by anxiety and St. Thomas Aquinas (1993, 1995, 2017) defined it in terms of a lack of joy and appetite for and interest in things.

In the Renaissance, writers like Petrarch (1991, 1992, 1999, 2016) and Dante (2014) alluded in their works to the lack of virtue and the inability to find hope or desire that defines boredom, that old sadness that torments both body and soul. Others, like St. Ignatius of Loyola (1992, 2014) and St. John of the Cross (2005) considered it to be a darkness of the soul or restlessness without hope, idleness or tepidity, bad mood or a physical malaise that results in displeasure, while Ficino (2019) spoke of a melancholic boredom leading to genius which, in excess, afflicts the soul with permanent tedium.

Great thinkers form the Enlightenment like Voltaire (1838, 2003, 2017) and Kant (2006; see also Ros Velasco, 2021) referred to boredom as a consequence of excess free time and a lack of work, a concept which the German romantic writer von Kleist (2003) would also later agree with. Along the same line, some of his contemporaries analyzed our social structure to approach boredom. Heine (2005, 2006, 2016) considered it to be a symptom of the inflexibility and oppression induced by society, while Tieck (1828, 1853, 1854) understood boredom to be a stagnation of the soul or a murky defect of life, a cultural exhaustion resulting from a distressing and exhausting psychic misery. Additionally, Büchner (1998) declared that boredom—the *Langeweile*—was an individual inability to make sense of our surroundings.

The French intellectuals of the nineteenth century, for their part, made an in-depth analysis of the experience of boredom in a host of philosophical literary works. For the Viscount of Chateaubriand (1952, 2014), boredom meant suffering from a serious paralysis of one's will. de Senancour (2010) expressed boredom as an opposition between what one imagines and what one experiences, a state in which one does not know what to wish for precisely because these wishes are disproportionate, and the imagination has promised too much. This is similar to what the authors Staël-Holstein (1835, 1995, 1998, 2000, 2010) and Sand (2022) advocated. A curious case is that of Flaubert (2003), the creator of one of the characters most afflicted by boredom of all time—Bovary. The novelist explicitly differentiated, for the first time in history, between two types of boredom (Flaubert, 1980): the banal tedium due to an excess of leisure time and the modern tedium of an existential nature—what we now know as simple or passing boredom, in contrast to complex or profound boredom. Baudelaire (2017) filled dozens of pages about this latter concept, describing it as desperation or the angst of hopelessness because everything one aspires to is unrealistic; the absence of desire is also what Tolstoy (2017) spoke about. Furthermore, other philosophers like Kierkegaard (2013) and Heidegger (1995) ascribed to the thought on this form of profound boredom that leads to angst and cognitive reevaluation, respectively.

Beyond philosophy and literature, many physiologists were interested in the state of boredom in the French context at the end of the eighteenth and nineteenth centuries. Hallé and Thillaye (1819) suggested that this was a case of a painful torment that brought with it considerable disarray; Esquirol (1821) appreciated it as a form of hatred of life; de Sauvages (1771) as a disease stemming from tiredness with existence; Littré (1863) as a vacuum of the soul deprived of an interest in things, while de Boismont (1850) referred to it as a moral disease of modern society. His understanding of boredom fully coincided with that of other philosophers with a sociological bias—the treatment of boredom from a purely sociological perspective has not begun to gain visibility until recently, thanks to works such as those by Conrad (1997), Gardiner and Haladyn (2016), or Ohlmeier et al. (2020), for example—like Simmel (1950, 1983), Kracauer (1995), Marx (2009), Lefebvre (2012, 2022), and Adorno and Horkheimer (2016), to mention but a few, for whom tedium was just a symptom of capitalism and its worst creatures—among them, the culture of mass entertainment. The philosopher Andreas Elpidorou (2014, 2017), now in contemporary times, has focused on the functional component of boredom as a sign of dissatisfaction with the present that alerts us to change something in our lives.

What interests all of us who dedicate ourselves to the study of boredom, whatever our discipline, is to prevent its pathological and dysfunctional instances. To make progress along these lines, it is essential to have contributions from philosophy and sociology on the different variables which condition the experience of boredom, such as its durability or its depth (Ros Velasco, 2024b). The study of boredom is a clear example of the fact that scientific rationality alone is insufficient to address a problem in its entirety and that it requires working together with the humanities to achieve an authentic understanding of the phenomenon under study. Through this unified effort, the anthropological, epistemological, ethical, and spiritual questions that determine the experience of boredom can be answered, beyond those such as psychophysiological questions, which are dealt with by scientific reasoning.

As far as the anthropological question is concerned, when studying boredom, the question of man and the idea that, in each historical and cultural period, underlies this construct cannot be left out. As beings living in society, it is undeniable that we are educated in certain values that determine how we understand and deal with boredom through free or conditioned decision-making, but also that we are genetically predetermined to suffer more or less boredom. On the epistemological question, expanded reason allows us to establish the boundaries of our respective branches of knowledge to explore the nature of boredom, making it easier for us to complement each other in order to fill the various gaps that arise from each of the disciplines that deal with it. It is impossible to shed light on this question without putting our respective approaches and methodologies to work together. The ethical question is essential, as we all want to know about boredom in order to guide our praxis, avoiding maladaptive responses to its suffering. In this regard, there is no point in knowing the neural circuits that underlie the experience of boredom if we do not then consider how this knowledge can be applied to improve our individual and social actions. This is what philosophy and theology have done throughout history when they have looked at boredom, as well as when they have studied it in order to answer the question of meaning in different metaphysical frameworks.

4 Blumenberg's Approach to Boredom from Expanded Reason

Although the question of boredom represents one of the most recurrent research topics in the scientific community today, many aspects of this reality are resistant to being apprehended from a scientific framework and that, therefore, we have to cover by making use of hypotheses and speculations closer to philosophy. These realities need, in spite of everything, to be understood in some way because, as explained above (see Sect. 2 in this chapter), it is only based on this understanding that we are able to achieve a certain horizon of meaning. Although they are reluctant to be pigeonholed within the objectivist pretensions of the human gaze, they are still part of a completely amorphous and disfigured experience that appears to us to be unfathomable, and which causes us impotence. We are forced, as Blumenberg did, to resort to symbolic forms that generate totality in order to obtain knowledge of the phenomenon applicable to everyday life.

Blumenberg has always argued that science has to be in dialogue with philosophy and theology in addressing a fundamental question that arises from our need to know more about human phenomena, and of course, he agreed to apply this premise to the study of an everyday phenomenon such as boredom. His approach to the question of boredom shows that knowledge about this state can never be the result of the isolated work of different disciplines, nor the juxtaposition of the philosophical and theological approach to the scientific, but of the integration of fragmentary knowledge into the whole. His approach to boredom from the perspective of expanded reason implies a reconciliation between scientific rationality and philosophical and theological reflection, doing justice to the philosophical, theological, literary, sociological, anthropological and medical narratives, among others, which have highlighted the interconnection between all of them and the need for an interdisciplinary approach from the perspective of expanded reason when dealing with a multifactorial phenomenon such as the one we are dealing with.

It is tremendously difficult to trace everything Blumenberg has written concerning boredom. Almost all the notions we have at hand are to be found in *Description of Man* (2011). A few loose mentions are to be found in works like *Life Time and World Time* (1986) o *Concepts in Stories* (2003). We also read something about melancholy in *Work on Myth* (1985). From these texts, we can deduce what the philosopher thinks about boredom and how his knowledge has drawn on countless disciplines. A complete reconstruction would necessarily lead us to his *Nachlass*, located in the DLA. Where we will not find an unpublished text on boredom, but dozens of them, small writings scattered in countless boxes and folders. Some of them are immediately obvious, just by entering the word *Langeweile* in the search engine of the DLA's Katalog Kallías. Others can be discovered because Blumenberg himself compiled a short index in his *Karteikarten* with the titles of several unpublished notes written on the subject of boredom.[1] But many of them are hidden under the most puzzling headings that would never lead one to suspect that in them Blumenberg is going to give us a key quotation to understand his approach to the question of boredom.[2] Almost none of the previous ones are dated. We know that the part of *Beschreibung des Menschen* (2011) in which he deals with boredom gathers manuscripts drawn up from the lectures he gave in the winter semester of 1976–1977 at the University of Munich. Several of the unpublished ones are in folders dated 1968–1988 and 1992–1993. When he wrote the rest is a mystery, but that is not what we are concerned with here either.

Blumenberg's study of boredom, combining philosophy with anthropology, literary and cultural studies with sociology, and palaeoanthropology with prehistory, all accompanied by a devastating critique of twenty-first century mental health sciences, leads him to define this state holistically as a painful emotion that we suffer whenever we are exposed to familiar situations or contexts that no longer represent a challenge for us, that is, in which we find ourselves in a situation of over-adaptation. What is essential for Blumenberg is not so much to analyze *why* boredom arises, but *what for*. The philosopher argues that boredom has the *function of* pointing out what is wrong, those situations that have no added value or meaning for us, to force us, through our desire to escape from pain, to make a change in the present. Blumenberg's approach is a real precedent for what some current philosophers have called the Functionality Theory of Boredom (its main representative is, among others, the aforementioned philosopher Andreas Elpidorou).

Blumenberg makes a strong point about how painful boredom is. In *Beschreibung des Menschen* (2011) he says that boredom makes us feel an irritation caused by an impasse self-awareness; it seems as if we are at a stalemate, as if nothing alluded us

[1] I will always be indebted to the researcher Pedro García-Durán, author of a chapter in this joint book, for introducing me to this index as soon as I set foot in the DLA. I also owe a very special thanks to Alberto Fragio Gistau for having been the instigator of my first visit to the archive.

[2] I only had a year to work on the unpublished Blumenberg writings, so I likely left a lot of material untouched among the *hidden writings*. Nor is it my intention, moreover, to dissect all those I managed to find in this chapter—they are all found in my doctoral thesis (Ros Velasco, 2017b), which I encourage the reader to read if he or she is hungry for more knowledge.

and as if we could not speak out on anything. In the unpublished text[3] "Langeweile, Kurzweil" (1992–1993) he explains that boredom can be experienced as a stillness similar to death, hence the use of expressions such as *dying of boredom* or references to *tödliche Langeweile* (deadly boredom) (Blumenberg, n.d.-m, UNF 373–375). Boredom has to hurt in order to fulfil its *function*, which is none other than to make us react in order to escape from those situations that are of insufficient value to us. He says in *Description* (2011) that the experience of boredom is so powerful that it cannot be ignored. Boredom, explains, "belongs to the strongest driving forces of human beings," being, at the beginning, like a crippling force that then results in "a force of violent repulsion" (Blumenberg, 2011, p. 530, a. trans.) in the situation that provokes it.[4] It cannot be otherwise if it is to make us react. In the end, boredom is nothing more than the result of the "poverty of stimulation and [the] prohibition of […] falling asleep" (Blumenberg, 2011, p. 530, a. trans.).

In this regard, the philosopher points to two moments of reaction that turn boredom into a functional experience. The first, which he announces in *Description* (2011), is merely a state of awareness of a situation that has become obsolete. It is a moment of awareness of the present context and of oneself within it. Later, the reaction involves the implementation or materialisation of some kind of escape strategy in order to implement a change that will free us from boredom. Blumenberg talks of boredom as "a state in which one do 'anything' just to escape of it" (1992–1993, Langeweile, Kurzweil, a. trans.). When facing the lack of stimulus that boredom represents, "internal compensation is generated as an invisible movement" (Blumenberg, 2011, p. 530, a. trans.). Blumenberg never says that the change to which boredom pushes us has to be unprecedented (see Ros Velasco, 2023); it is enough for him that such a movement takes place.[5]

Blumenberg understood boredom's functionality in an *adaptive* sense. Kant said in his *Anthropologie* (2006) that creatures naturally tend to seek comfort and a state of optimal adaptation. However, as Blumenberg explains in *Description of Man* (2011), an excess of adaptation is not desirable either, because it results in the stillness that undermines our ability to react and prevents us from adapting to sudden changes. The task of boredom is none other than to prevent stagnation, and too much stillness. For Blumenberg, we must not stay in a situation we are already familiar with, i.e. that makes us bored, because only then are we impelled to introduce change and keep ourselves in the constant motion that greases our capacity to adapt to the unknown that may be coming. Pain pushes us to escape and alter our

[3] Unpublished manuscripts by Blumenberg in this chapter are already in my doctoral thesis (Ros Velasco, 2017b), having been got the DLA and Bettina Blumenberg's permissions for their reproduction.

[4] These quotes are from an untitled and unnumbered card in the box *Zettelkasten 26 U-Welt* at DLA.

[5] In my opinion, if in spite of feeling the pain, and even *thinking about* the pain, it is not possible to complete the circle of reaction to which Blumenberg refers, boredom cannot be called functional, because it does not fulfil the ultimate function it is supposed to, which is to force us to introduce change into the context. Functional, *sensu stricto*, is only transient situational boredom. This will be seen more clearly later.

position (Blumenberg, 1968–1988, 019434), allows us to reject situations that have become too comfortable and provokes the search for different experiences, avoiding stagnation and the reduction of our adaptive capacities in the face of unexpected circumstances. Boredom remains when we are confronted with experiences that do not live up to our expectations because they are too familiar, preventing us from enjoying them forever. This makes it difficult for us to remain in our *comfort zone* forever, which, would be a dangerous stagnation, an oxidation of our adaptive mechanisms, urging us to *reoccupy* different spaces.

For Blumenberg, the important message is to keep moving, whatever the direction, even if it leads to maladaptive or seemingly harmful responses. Blumenberg himself is well aware that there are many cases in which boredom ends up being the starting point for disaster. Nevertheless, he finds the chaos that may result from the attempted escape from boredom preferable, functionally, to the chaos we would face if we were never bored. He is clear that not having the mechanism that underlies boredom would lead us to a real catastrophe, of much greater magnitude than those that, throughout history, have been unleashed because of it. He knows that boredom will not always lead to constructive change, and that in many cases change will be destructive, but he finds immobility more destructive. Without boredom in the face of the familiar, we would become so comfortable and over-adapted that we would become extinct at the slightest change. Blumenberg does not hesitate to point out that if boredom does not appear at the moment we are fully adapted, forcing us to be in constant motion, we would be unable to adapt to future major changes. For this reason, Blumenberg celebrates that boredom wakes up "the spirit of adventure" (1992–1993, 019297–019298, a. trans.), that it represents "some kind of willingness to be able to do anything" (n.d.-o, UNF 820–821, a. trans.), that it is, according to his view, "the entrance door for what is not in accordance with the order, for what is contrary to the system" (2011, p. 536, a. trans.).

Blumenberg analyzed many strayings in his time caused by attempts to escape boredom, an "annoying boredom" (2011, p. 534, a. trans.), also called profound boredom, a consequence of classical contemporaneity, shared by many, resulting from certain socio-cultural structures or constructs that were as jaded as they were constrictive and immovable, and which limited human creativity in responding to the tedium they released, making it impossible to implement a collective escape/cope strategy from the *source of boredom* itself (Ros Velasco, 2025b). Blumenberg's society is a *mass* that is bored by an apparently immovable situation that becomes chronic and which, over time, casts doubt on the meaning of everything, endangering the optimism founded by the idea of progress and by the science on which all had placed their gaze and their trust at the very beginning of the past century. The boredom of Blumenberg's time, in his own words, is the child of the "new asceticism" and its ideal of progress, of a *moralia* that leads to "living without living" (2003, p. 32, a. trans.) by constantly seeking novelty to deal with tedium (n.d.-d, 0219685).

Blumenberg saw how the struggle against *ennui* in this period reinforced the taste for the repulsive, for illness, for animalism, for primitivism, for infantilism, for bestiality, for violence and extreme barbarism, for morbidity and lasciviousness

(1968–1988, 019424). Mass boredom favored the compulsive and excessive action typical of "actionisms" (1992–1993, 019297–019298, a. trans.), as a 'lack' that must be compensated by an amplification (1992–1993, Unbehagen) that puts on the table our "more intimate desires of death" (Blumenberg, 2011, p. 539, a. trans.). In many cases, Blumenberg says, "the aesthetic nature of the revolts against boredom" (2011, p. 537, a. trans.) hinted, for example, at a macabre taste for war. Blumenberg was not at all surprised that Valéry and Gide exchanged letters in which they expressed a yearning for battle (see Ros Velasco, 2022, 2024a). Both were even envious of the soldiers who, during the events of 1 May 1891, had killed ten demonstrators against the last European war of the nineteenth century in order to stave off boredom. On the evening of 8 May 1891, Valéry wrote to Gide: "I almost wish for a monstrous war to run away with [...] The days are a yawn of boredom [...] Does this barbarian surprise you?," to which Gide replied, with great complicity, on the morning of 12 May 1891, "Is it you, this warrior? Ah! [...] you too dream of the trembling of weapons and blood" (1955, pp. 82–83, a. trans.). Nor was he frightened by the fact that the playwright Friedrich Hebbel asked himself "Why should not a man commit murder just to avoid boredom?" (n.d.-m, UNF 373–375, a. trans.), nor when he knew about the IWW decorated combatant Berthold Feuchtwanger's confession that many of his exploits had been motivated only by bets he had made with the Superior Officer to kill boredom (n.d.-j, UNF 2348). Then, the "apocalyptic desire for the great war" was completely normalized (Blumenberg, 2011, p. 538, a. trans.). Blumenberg expressed that the general rejection of boredom "caused more victims than hatred" (n.d.-j, UNF 2348, a. trans.), although, sometimes, both factors would have gone hand in hand, as in the case of the figure of Adolf Hitler, whom his doctor, Theo Morell, classified among type A patients who suffered from deep and severe boredom (n.d.-h, UNF 1835–1839). For Blumenberg (2011), boredom was probably not only behind the Nazi genocide, but any act of terrorism. Blumenberg was aware that boredom had reached in his time the status of a plague as "cholera and pest, leprosy and smallpox" (n.d.-g, UNF 166–167, a. trans.) against which the only cure was extravagance, deviance, and illness. But all this debauchery was preferable to boredom: "a 'little bit of hysteria' makes the world more colorful than pedantic boredom" (Blumenberg, n.d.-e, Hysterie, a. trans.), he exclaimed in another unpublished piece.

Blumenberg points out, in *Beschreibung des Menschen*, that "boredom can be the metaphysical principle of the world" (2011, p. 529, a. trans.), of creation, while recognizing, in the unpublished "Umkehrung eines Mythos" (n.d.-k, UNF 2461), that "also the end of the world may be motivated by the same reason, the boredom with it [the creation]" (a. trans.). Perhaps we have begun to inhabit a time when boredom is more likely to be experienced destructively than constructively. However, it was not always so. Blumenberg is convinced that boredom played an essential role in the past, in our evolution as a species, that the possibility of being bored represented a real opportunity for growth, even an evolutionary advantage, in our most remote ancestors, at times when they found themselves over-adapted, because the fact of wanting to abandon the pain of what was too familiar encouraged them to introduce change in their context without which we would not be where we are. Boredom was the prelude to culture.

Blumenberg (2011) explains that boredom is not an essential emotion in human beings, but a gained one by our ancestors since it was functional for self-preservation. It is "a unitary form of anticipatory behavior adjusted to initial anthropological situations" (2011, p. 524, a. trans.). Contemporary thinkers as Professor Peter Toohey understand boredom in this sense, as an adaptive emotion "in a true Darwinian sense" (2011, p. 7), that has been playing an unavoidable function since time immemorial. The majority, however, believe that boredom can only be found in social structures that generate enormous amounts of free time and that, consequently, in earlier times our predecessors were too busy with their own survival to have time to be bored, which is as false as it is fallacious (Ros Velasco, 2022, 2023, 2024a). Blumenberg has no doubt that our prehistoric relatives were bored, and although this cannot be proved, since boredom does not fossilise, Blumenberg applies the principle of insufficient reason here and has no problem in embracing "the rationality of the absence of substantiation" and "allowing certain degrees of indeterminacy" (2013, pp. 98, 13, a. trans.; see Sect. 2 in this chapter), typical of the approach to phenomena from the methodology of expanded reason, to get to the bottom of the matter as far as the essence of boredom is concerned.

Blumenberg thinks that boredom began being experienced when the *Homo sapiens*' relatives were adapted enough, likely when they felt themselves safe and confident inside the caves, going unnoticed by predators. As he recounts on a card, "the cave became a crucial alternative to the jungle" (Blumenberg, n.d.-b, 020141, a. trans.). It was the place where "the inestimable advantage of undisturbed reproduction" (Blumenberg, n.d.-a, 019682, a. trans.) and "safe sleeping accommodation" (Blumenberg, 1968–1988, 022198, a. trans.) were assured. "In the firm grip of the homey cave" (1968–1988, 17829, a. trans.), Blumenberg says, our ancestors found themselves protected from threats; continuous monitoring was no longer necessary, giving way to states which would have been impossible to think about during periods of life outdoors, such as relaxation, deep sleep, and boredom. The cave, as well as other types of shelters, brought tranquillity and greater adaptation to the environment for our ancestors, to the extent that they could begin to experience some degree of boredom. Blumenberg takes it for granted when, in his narrations about the daily scenes of our ancestors, he explains that, sheltered in the caves, they began to tell stories and sing songs to relieve "boredom of the dark nights" (n.d.-c, 021606, a. trans.).

For Blumenberg, those more prone to get bored were precisely those who spent more time sheltered in shelters, that is to say, mothers, children, elders, and wounded, those who, having remained unfit for physical tasks such as hunting for being considered the *weakest*, lived their lives in a more monotonous and less exciting way. The forays into these caves would have helped them to survive, but, at the same time, they would have encouraged the appearance of a nagging boredom so that they would not become too comfortable. At a time when they did not have to face as many problems and conflicts as in the past, and when survival was not a constant necessity, boredom could have been a factor. The use of fire in these camps would not have been a minor detail for Blumenberg. It would have made it possible to leave behind "the lack of light in which mankind had spent most of its history" (Blumenberg, n.d.-i, UNF 2271; see also Blumenberg, 2002, pp. 110–112, a. trans.) and overcome the tragedy of nights whose "length exceeded the organic need for sleep" (Blumenberg, n.d.-q, UNF 963, a. trans.). But then, the new daylight hours would also have had to be spent on something. It has been widely argued in literature that this activity must have been socialization (see Ros Velasco, 2017b). The nights would be longer, the waking hours would increase, and there would be more time to engage in other activities, such as storytelling to ward off boredom (Blumenberg, n.d.-i, UNF 2271).

Blumenberg argues that the *weak engaged* in activities like this to remedy boredom, sowing the seeds of a proto-culture. But, furthermore, Blumenberg says, boredom would have led the *weakest* to do that "for which the others had neither time nor will – they made life worthwhile beyond survival" (n.d.-c, 021606, a. trans.). Stories and songs against boredom brought "a powerful joy" (Blumenberg, n.d.-c, 021606, a. trans.) to the shelters, with which the *weak* also compensated for their *weakness* and inverted the social structure known until then. The division of the group into *weak* and *strong* had been "the big problem of humanity" (Blumenberg, n.d.-c, 021606, a. trans.). Thanks to hunting "the powerful and the wily benefited from their position in the group to took distributive advantage over the others" (1968–1988, 019592, a. trans.). However, in the situation of the *weakest*, "the big, mortal boredom is spreading, pushing the new 'madmen' into random positions as only formally simulated elites" (Blumenberg, n.d.-l, UNF 320–322, a. trans.). Determined to cut off boredom caused by "the protection offered by a stable way of life," by "the always identical," they ran the risk of taking a step on their own even if the elites called them 'madmen' (Blumenberg, n.d.-l, UNF 320–322, a. trans.). "They began to sing and spin around [the strongest]" (Blumenberg, n.d.-c, 021606, a. trans.). In the end, Blumenberg states, the *weakest* were the "essence and soil of every culture" (n.d.-c, 021606, a. trans.), those that led to the creation of "abstract or technical systems" (n.d.-f, Tiger-Fox, 238, a. trans.) and proto-culture. Thus, the bored-weak could take part in the evolutionary line because they developed certain skills in their effort to avoid boredom inside the cave "which allow them to win preference in the selection" (Blumenberg, n.d.-d, 0219685, a. trans.) over those who tried to impose force. From then on, the tables would turn: the elite members would have to compete with the hitherto marginalised for the favour of the females, who would value rational qualities more than strength. From then on, those who could hold better conversations, tell better stories and, in short, provide better company to relieve boredom, would be the most valued in the clans.

To sum, those ancestors "overcame boredom and mental powers were increased" (Blumenberg, 1968–1988, 021851, a. trans.); "they were elevated" (Blumenberg, n.d.-n, UNF 472–476, a. trans.), thanks to boredom, towards a way of life that strived to transcend, that went beyond survival. Boredom, or rather the reaction to boredom, may have pushed them towards the breakdown of the perfectly regulated relationship with the environment, without which, perhaps, "the human being would not exist" as he is now (Blumenberg, 2013, p. 67, a. trans.). From the philosopher's notes comes the suggestion that boredom was present in our prehistory, at a time of great stability, and that in our ancestors' effort to find relief from it, there was a response that eventually led to a paradigm shift. Nietzsche said, in *The Antichrist* (1999), that all the myths, songs, dances and inventions we have created since the beginning of human history respond to boredom. That may be so, but it is also possible that these cultural manifestations originated to pass down group wisdom between generations, just as it is also possible that, in the beginning, they were born only as entertainment, but that their potential for cohesion, prosperity and the survival of peoples became manifest later on. We cannot know. This is also true of other proto-cultural manifestations which, unlike the previous ones, remain in time. For example, it is plausible that prehistoric artistic production was simply a way of filling time for those ancestors who had reached higher cognitive development and higher levels of adaptation.

The multidisciplinary and interdisciplinary knowledge regarding the phenomenon of boredom that Blumenberg acquired through his expanded reason approach was the basis for the unprecedented insights he was able to postulate about its nature that led him to place great emphasis on how important it is to learn to *tolerate* boredom. This means, no more and no less, that we should not try to eliminate boredom by taking the easy way out, as mental health sciences sometimes propose, but become interested observers of what it has to say (Blumenberg, n.d.-p, UNF 935–937). Blumenberg was against attributing to boredom the status of pathology or disease to be dealt with immediately. He said in *Beschreibung* (2011) that the WHO had turned everything into pathology with its definition of health as *the complete physical, mental and social well-being*, promoviendo que "everything becomes a disease and […] [that] the need for treatment is a permanent issue" (2011, p. 516, a. trans.). Blumenberg did not want boredom to become an annoyance from which health professionals can earn a living imposing their "therapeutic claims" (2011, p. 520, a. trans.). He did not want boredom, or our most destructive reactions to boredom, to end up giving place to what he called "a diagnosis with promising future" (2011, p. 527, a. trans.) to which "a very harsh theraphy" (2011, p. 551, a. trans.) would have to be applied, forever losing sight of the part of responsibility that the context has in its suffering.

Currently, boredom has already become a business. Advertisers and the media make use of boredom as a 'sales concept,' while no nineteenth-century journalist talked about boredom (Blumenberg, n.d.-g, UNF 166–167). Blumenberg feared that it was only a matter of time before the more licentious and authoritarian medicine—the kind that made Foucault's hair stand on end—would take advantage of the more complex experiences of boredom. This, at the moment, has not yet happened. According to the Blumenbergian diagnosis, perhaps reversion is still possible. Boredom may still be understood as some kind of deviation, not necessarily pathological but beneficial.

5 Conclusion

Blumenberg's application of expanded reason to the study of boredom shows that, nowadays, the approach to this phenomenon from the point of view of scientific rationality has become stagnant by disregarding the whole philosophical, literary and theological tradition that had developed around it in previous centuries. This has happened because the question of boredom has been approached solely from the standpoint of scientific rationality which, on its own, is not capable of responding to the questions that have to do with man, with the good, with truth and with meaning, and which are precisely those that delimit the experience of boredom and give us a clue as to what the real *raison d'être* of this emotion is and how its function has unfolded throughout history.

In recent decades, the humanities reclaimed their inescapable role in the study of boredom after almost a century in which it has only been discussed in the mental health disciplines. However, humanists are often producing work that makes the same mistake that was made for much of the last century by the mental health sciences: leaving one unknown out of the equation. Blumenberg's approach is so special because, by gathering the knowledge at his disposal from a wide range of disciplines (philosophy, literature, anthropology, sociology, psychology, etc.), he was able to develop a whole original thesis on the experience of boredom which, in many cases, subsequent researchers *reproduce*, ignoring that what they are talking about is part of Blumenberg's harvest.

Blumenberg demonstrates the interdependence between the presuppositions of the mental health sciences and those of the humanities—contributed by philosophers and theologians of all times, among other intellectuals—by presenting the existing links between the different narratives, as well as showing the limits of scientific rationality. In a society that increasingly works in isolation, in which interdisciplinarity and multidisciplinarity are penalised as a lack of specialization, the scholar Blumenberg encourages us to work from the perspective of expanded reason, to reconciliation that allows us to read each other, listen to each other and collaborate in cross-projects, that is, enable the continuity of the synthesis of knowledge that facilitated in the past the goal of reaching a complete understanding of what the experience of boredom represents from the perspective of expanded reason and of guiding human praxis in responding adaptively to its suffering. His work serves as an example and inspiration to researchers seeking to elucidate aspects of human nature that will enable us not only to expand our knowledge, but also to give meaning to our existence and guide our praxis, so that we dare to cross the boundaries of our respective sciences and explore the benefits of the application of expanded reason. Research of the future will be done together or not at all.

References

Adorno, T. W., & Horkheimer, M. (2016). *Dialectic of enlightenment*. Verso Books.
Barmack, J. E. (1939a). A definition of boredom. *American Journal of Psychology, 52*(3), 467–471.
Barmack, J. E. (1939b). Studies on the psychophysiology of boredom: Part I. *Journal of Experimental Psychology, 25*(5), 494–505.
Baudelaire, C. (2017). *Delphi collected poetical works of Charles Baudelaire*. Delphi Classics.
Bergler, E. (1945). On the disease-entity boredom ("Alyosis") and its psychopathology. *Psychiatric Quarterly, 19*(1), 38–51.
Berlyne, D. E. (1960). *Conflict, arousal and curiosity*. McGraw-Hill.
Berlyne, D. E. (1967). Arousal and reinforcement. In D. Levine (Ed.), *Nebraska symposium on motivation* (Vol. 15, pp. 1–110). University of Nebraska Press.
Bieber, I. (1951). Pathological boredom and inertia. *American Journal of Psychotherapy, 5*(2), 215–225.
Blumenberg, H. (1968–1988). 019424. Musil, Tagebücher Heft 30: Etwa März 1929 bis November 1941 und später (I 801). In *Zettelkasten 01: Anthropologie*. DLA.
Blumenberg, H. (1968–1988). 019434. In *Zettelkasten 01: Anthropologie*. DLA.
Blumenberg, H. (1968–1988). 019592. Alle Tierpopulationen haben Ökologie, nur die menschlichen Populationen haben Ökonomie. In *Zettelkasten 01: Anthropologie*. DLA.
Blumenberg, H. (1968–1988). 021851. In *Zettelkasten 01: Anthropologie*. DLA.
Blumenberg, H. (1968–1988). 022198. Übersprungeinschlafen (Bilz, Paläoanthr. 22) Resignationsschlaf. In *Zettelkasten 01: Anthropologie*. DLA.
Blumenberg, H. (1968–1988). 17829. In *Zettelkasten 01: Anthropologie*. DLA.
Blumenberg, H. (1985). *Work on myth*. Penguin Random House.
Blumenberg, H. (1986). *Lebenszeit und Weltzeit*. Suhrkamp.
Blumenberg, H. (1992–1993). 019297–019298. Zeitgewinn und Langeweile. In *Konvolut text und Materialsammlung Langeweile*. DLA.
Blumenberg, H. (1992–1993). Langeweile, Kurzweil. In *Konvolut text und Materialsammlung Langeweile*. DLA.
Blumenberg, H. (1992–1993). Unbehagen: Land der Langeweile. In *Konvolut text und Materialsammlung Langeweile*. DLA.
Blumenberg, H. (2002). *Vor allem Fontane. Glossen zu einem Klassiker insel taschenbuch*. Insel.
Blumenberg, H. (2003). *Conceptos en historias*. Síntesis.
Blumenberg, H. (2010). *Paradigms for a Metaphorology*. Cornell University Press.
Blumenberg, H. (2011). *Descripción del ser humano*. FCE.
Blumenberg, H. (2013). *Teoría del mundo de la vida*. FCE.
Blumenberg, H. (n.d.-a). 019682. Biotopwechsel als "genetische Drift." In *Zettelkasten 03: Deutsche Aufklärer*. DLA.
Blumenberg, H. (n.d.-b). 020141. Vorhersage möglicher Fossilfundstätten im bereich junger geologischer auffaltungen. In *Zettelkasten 03: Deutsche Aufklärer*. DLA.
Blumenberg, H. (n.d.-c). 021606. Die Koalition der Schwachen und die Sklaverei der Eliten. In *Zettelkasten 03: Deutsche Aufklärer.* DLA.
Blumenberg, H. (n.d.-d). 0219685. Ist der Kampf ums Dasein eine Metapher? In *Zettelkasten 03: Deutsche Aufklärer*. DLA.
Blumenberg, H. (n.d.-e). Hysterie. Ein fragwürdiger Begriff. In *Konvolut Materialsammlung Anthropologie 3/3*. DLA.
Blumenberg, H. (n.d.-f). Tiger-Fox, 238. In *Zettelkasten 03: Deutsche Aufklärer*. DLA.
Blumenberg, H. (n.d.-g). UNF 166-167. Diesseits der Langeweile. In *Schuber 1, Mappe 2*. DLA.
Blumenberg, H. (n.d.-h). UNF 1835–1839. Hitlers Bakterien. In *Schuber 7, Mappe 4*. DLA.
Blumenberg, H. (n.d.-i). UNF 2271. Kindeswohl und Lichtbedürfnis. In *Schuber 5, Mappe 6*. DLA.
Blumenberg, H. (n.d.-j). UNF 2348. Die Indifferenzen der Langeweile. In *Schuber 5, Mappe 7*. DLA.
Blumenberg, H. (n.d.-k). UNF 2461. Umkehrung eines Mythos. In *Schuber 6, Mappe 3*. DLA.
Blumenberg, H. (n.d.-l). UNF 320–322. Todestrieb der Eliten. In *Schuber 1, Mappe 3*. DLA.

Blumenberg, H. (n.d.-m). UNF 373–375. Tödliche Langeweile. In *Schuber 1, Mappe 3*. DLA.
Blumenberg, H. (n.d.-n). UNF 472–476. Rache der verstoβenen Metaphysik. In *Schuber 2, Mappe 1*. DLA.
Blumenberg, H. (n.d.-o). UNF 820–821. Hohe Ziele und mindere Zwecke. In *Schuber 2, Mappe 4*. DLA.
Blumenberg, H. (n.d.-p). UNF 935–937. Langeweile. In *Schuber 3, Mappe 1*. DLA.
Blumenberg, H. (n.d.-q). UNF 963. Künstliche Beleuchtung. In *Schuber 3, Mappe 2*. DLA.
Büchner, G. (1998). *Danton's death; Leonce and Lena; Woyzeck*. Oxford University Press.
Campbell, R. N. (1996). *Psychiatric dictionary*. Oxford University Press.
Cassian, J. (1999). *The monastic institutes: On the training of a monk and the eight deadly sins*. Saint Austin Press.
Cassian, J. (2015). *The collations*. Gracewing Publishing.
Cassirer, E. (1965). *The philosophy of symbolic forms*. Yale University Press.
Chateaubriand, F.-R. (1952). *Atala and Rene*. University of California Press.
Chateaubriand, F.-R. (2014). *Memoirs from beyond the tomb*. Penguin Random House.
Conrad, P. (1997). It's boring: Notes on the meanings of boredom in everyday life. *Qualitative Sociology, 20*, 465–475. https://doi.org/10.1023/A:1024747820595
Cooper, R. (1968). Psychology of boredom. *Ekistics, 25*(151), 404–408.
Corsini, R. J. (1999). *The dictionary of psychology*. Psychology Press.
Danckert, J., Hammerschmidt, T., Marty-Dugas, J., & Smilek, D. (2018). Boredom: Under-aroused and restless. *Consciousness and Cognition, 61*, 24–37. https://doi.org/10.1016/j.concog.2018.03.014
Dante, A. (2014). *The divine comedy: Inferno, Purgatorio, Paradiso*. Penguin Random House.
de Boismont, A. B. (1850). *De l'ennui*. Tædium vitæ, *vol.* 2. Martinet.
de Sauvages, B. (1771). *Nosologie méthodique, vol.* 2. Chez Hérissant le Fils.
de Senancour, É. P. (2010). *Obermann*. Noumena Press.
Elpidorou, A. (2014). The bright side of boredom. *Frontiers in Psychology, 5*, 1245. https://doi.org/10.3389/fpsyg.2014.01245
Elpidorou, A. (2017). The good of boredom. *Philosophical Psychology, 31*(3), 323–351. https://doi.org/10.1080/09515089.2017.1346240
Esquirol, J. É. D. (1821). Suicide. In *Dictionnaire des sciences médicales, par une société de médecins et de chirurgiens* (pp. 213–283). Panckoucke.
Farmer, R., & Sundberg, N. D. (1986). Boredom proneness: The development and correlates of a new scale. *Journal of Personality Assessment, 50*(1), 4–17. https://doi.org/10.1207/s15327752jpa5001_2
Ficino, M. (2019). *Three books on life*. University of Chicago Press.
Flaubert, G. (1980). *The letters of Gustave Flaubert: 1830–1857* . Harvard University Press.
Flaubert, G. (2003). *Madame Bovary*. Penguin Random House.
Gardiner, M., & Haladyn, J. J. (2016). *Boredom studies reader: Frameworks and perspectives*. Routledge.
Gide, A., & Valéry, P. (1955). *Correspondance 1890–1942*. Gallimard.
Greenson, R. R. (1951). Apathetic and agitated boredom. *Psychoanalytic Quarterly, 20*, 346–347.
Greenson, R. R. (1953). On boredom. *Journal of the American Psychoanalytic Association, 1*, 7–21. https://doi.org/10.1177/000306515300100102
Hallé, J.-N., & Thillaye, J.-B.-J. (1819). *Dictionnaire des sciences médicales, par une société de médecins et de chirurgiens*. Panckoucke.
Heidegger, M. (1995). *The fundamental concepts of metaphysics: World, finitude*. Indiana University Press.
Heine, H. (2005). *Pictures of travel*. University of Michigan Press.
Heine, H. (2006). *Book of songs*. University of Michigan Press.
Heine, H. (2016). *The works of Heinrich Heine, VII. French affairs: Letters from Paris*. Leopold Classic Library.
Horace. (2011). *Satires and epistles*. Oxford University Press.

Hozier, A. Q. (1988). Functions of boredom: Treatment implications. *The Psychotherapy Patient, 3*(3–4), 43–54.
Instituto de Razón Abierta (2024). *Premios Razón Abierta* [website]. Instituto de Razón Abierta. https://razonabierta.org/premios-razon-abierta/
Kant, I. (1998). *Critique of pure reason*. Cambridge University Press.
Kant, I. (2006). *Anthropology from a pragmatic point of view*. Cambridge University Press.
Kant, I. (2007). *Anthropology*. Cambridge University Press.
Kierkegaard, S. (2013). *Kierkegaard's writings IV, Part II: Either/Or*. Princeton University Press.
Kracauer, S. (1995). *The mass ornament: Weimar essays*. Harvard University Press.
Lefebvre, H. (2012). *Introduction to modernity*. Verso Books.
Lefebvre, H. (2022). *On the rural: Economy, sociology*. University of Minnesota Press.
Levinger, E. (1951). Von der menschlichen Langeweile. Psychologisch-biologische Betrachtungen. *Schweizerische Medizinische Wochenschrift, 81*(5), 113–115.
Lipps, T. (1909). *Leitfaden der Psychologie*. Wilhelm Engelmann.
Littré, É. (1863). *Dictionnaire de la langue français*. Hachette.
Lucretius. (2008). *De Rerum Natura*. University of Wisconsin Press.
Maggini, C., & Dalle Luche, R. (1987). Per una psicopatologia della noia. Alcuni richiami storici alle relazioni tra noia e melanconia. *Rivista Sperimentale di Freniatria, 111*(5), 43–66.
Marx, K. (2009). *The economic and philosophic manuscripts of 1844 and the communist manifesto*. Prometheus Books.
Merker, B. (2013). Geschichte(n) der Paläoanthropologie. In C. Borck (Ed.), *Blumenberg beobachtet. Wissenschaft und Philosophie* (pp. 111–125). Karl Alber Verlag.
Münsterberg, H. (1913). *Psychology and industrial efficiency*. Houghton-Miffin.
Nietzsche, F. (1999). *Der Antichrist* (Vol. 11). Goldmann.
O'Hanlon, J. E. (1981). Boredom: Practical consequences and a theory. *Acta Psychologies, 49*(1), 53–82. https://doi.org/10.1016/0001-6918(81)90033-0
Ohlmeier, S., Finkielsztein, M., & Pfaff, H. (2020). Why we are bored: Towards a sociological approach to boredom. *Sociological Spectrum, 40*, 208–225. https://doi.org/10.1080/02732173.2020.1753134
Patrick, G. (1982). Clinical treatment of boredom. *Therapeutic Recreation Journal, 16*, 7–12.
Petrarch, F. (1991). *Remedies for fortune fair and foul*. Indiana University Press.
Petrarch, F. (1992). *Letters of old age*. The Johns Hopkins University Press.
Petrarch, F. (1999). *Canzoniere*. Indiana University Press.
Petrarch, F. (2016). *My secret book*. Harvard University Press.
Ponticus, E. (2006). *The Praktikos: Chapters on prayer*. Cistercian Studies.
Pope Benedict XVI (2006a, 12 September*). Lecture of the holy father: Faith, reason and the university. Memories and reflections*. Aula Magna of the University of Regensburg. https://www.vatican.va/content/benedict-xvi/en/speeches/2006/september/documents/hf_ben-xvi_spe_20060912_university-regensburg.html
Pope Benedict XVI (2006b, 19 October). *Address of his holiness Benedict XVI to the participants of the Fourth National Ecclesial Convention*. Verona Exhibition Centre. https://www.vatican.va/content/benedict-xvi/en/speeches/2006/october/documents/hf_ben-xvi_spe_20061019_convegno-verona.html
Revers, W. J. (1983). Die Langeweile—Symptom emotionaler Verkümmerung. *Zeitschrift für Klinische Psychologie, Psychopathologie und Psychotherapie, 31*(1), 4–13.
Ros Velasco, J. (2012). Metaforología y antropología en Hans Blumenberg. *Azafea. Revista de Filosofía de la Universidad de Salamanca, 14*, 207–231.
Ros Velasco, J. (2017a). Lo que no nos cuenta el método científico. Una revisión desde la filosofía blumenberguiana. In A. Campillo Meseguer & D. Manzanero Fernández (Eds.), *Las fronteras de la humanidad* (pp. 9–24). Red Española de Filosofía.
Ros Velasco, J. (2017b). *El aburrimiento como presión selectiva en Hans Blumenberg*. PhD Dissertation. Complutense University of Madrid.

Ros Velasco, J. (2017c). Boredom: A comprehensive study of the state of Affaris. *Thémata. Revista de Filosofía, 56*, 171–198. https://doi.org/10.12795/themata.2017.i56.08
Ros Velasco, J. (2019). Hans Blumenberg's philosophical anthropology of boredom. In A. Fragio Gistau, M. Philippi, & J. Ros Velasco (Eds.), *Metaphorologie, Anthropologie, Phänomenologie. Neue Forschungen zum Nachlass Hans Blumenbergs* (pp. 91–107). Karl Alber Verlag.
Ros Velasco, J. (2021). Kantian lange weile within the contemporary psychology of boredom. In M. Failla & N. Sánchez Madrid (Eds.), *Kant on emotions: Critical essays in the contemporary context* (pp. 171–184). Walter de Gruyter.
Ros Velasco, J. (2022). *La enfermedad del aburrimiento*. Alianza.
Ros Velasco, J. (2023). Contemporary myths on boredom. *Frontiers in Sociology, 8*. https://doi.org/10.3389/fsoc.2023.1183875
Ros Velasco, J. (2024a). *The disease of boredom*. Princeton University Press.
Ros Velasco, J. (2024b). Brief history of psychology of boredom. In M. Bieleke, W. Wolff, & C. Martarelli (Eds.), *The Routledge international handbook of boredom* (pp. 106–130). Routledge.
Ros Velasco, J. (2025a). Filosofía y antropología del aburrimiento en Hans Blumenberg. In J. Ros Velasco, E. J. Torregroza Lara, & O. A. Quintero Ocampo (Eds.). Nexofia.
Ros Velasco, J. (2025b). The experience of boredom in classical contemporaneity. In J. Ros Velasco & A. Elpidorou (Eds.), *Boredom in rewriting the history of philosophy*. Routledge.
Sand, G. (2022). *Delphi collected works of George Sand*. Delphi Classics.
Savitz, J., & Friedman, M. I. (1981). Diagnosing boredom and confusion. *Nursing Research, 30*(1), 16–20.
Seneca. (2004). *On the shortness of life*. Penguin Random House.
Seneca. (2009). *Dialogues and essays*. Oxford University Press.
Seneca. (2014). *Moral letters to Lucilius*. Aegitas.
Simmel, G. (1950). *The sociology of Georg Simmel*. Simon and Schuster.
Simmel, G. (1983). *Einleitung in die Moralwissenschaft: Eine Kritik der ethischen Grundbegriffe* (Vol. 1). Scientia.
Sommer, D., Ros Velasco, J., & Abarca, M. (2021). Bored: A pandemic of domestic violence. *Partner Abuse, 12*(1), 80–93. https://doi.org/10.1891/PA-2020-0022
St. Augustine of Hippo. (2006). *Instructing beginners in faith*. New City Press.
St. Augustine of Hippo. (2007). *Essential sermons*. New City Press.
St. Augustine of Hippo. (2008). *Confessions*. Oxford University Press.
St. Ignatius of Loyola. (1992). *The spiritual exercises of Saint Ignatius*. Loyola University Press.
St. Ignatius of Loyola. (2014). *A Pilgrim's journey: The autobiography of St. Ignatius of Loyola*. Ignatius Press.
St. John of the Cross. (2005). *The dark night of the soul*. Barnes and Noble Publishing.
St. Thomas Aquinas. (1993). *Commentary on Aristotle's Nicomachean ethics*. St. Augustine Press.
St. Thomas Aquinas. (1995). *On evil: Disputed questions*. University of Notre Dame Press.
St. Thomas Aquinas. (2017). *Summa Theologica: Part II-II (Secunda Secundae)*. Pinnacle Press.
Staël-Holstein, A. -L. G. (1835). *The influence of literature upon society*. Pearson.
Staël-Holstein, A. -L. G. (1995). *Delphine*. Northern Illinois University Press.
Staël-Holstein, A. -L. G. (1998). *Corinne, or Italy*. Oxford University Press.
Staël-Holstein, A. -L. G. (2000). *Ten years of exile*. Northern Illinois University Press.
Staël-Holstein, A. -L. G. (2010). *Reflections on suicide*. Kessinger Publishing.
Stuart, M. (1951). Boredom in a case of depression. *Psychiatric Quarterly, 20*, 346–349.
Sundberg, N. D., & Bisno, H. (1983). *Boredom at life transitions—Adolescence and old age*. Meeting of the Western Psychological Association.
Thakurdas, H., & Thakurdas, L. (1979). *Dictionary of psychiatry*. MTP.
Tieck, L. (1828). William Lowell. In *Schriften* (Vol. 6). Georg Reimer.
Tieck, L. (1853). Abendgespräche. In *Schriften* (Vol. 25). Georg Reimer.
Tieck, L. (1854). Die Vogelscheuche. In *Schriften* (Vol. 11). Georg Reimer.
Tolstoy, L. (2017). *Anna Karenina*. Penguin Random House.

Toohey, P. (2011). *Boredom: A lively history*. Yale University Press.
VandenBos, G. R. (2007). *APA dictionary of psychology*. APA.
Vogel-Walcutt, J., Fiorella, L., Carper, T., & Schatz, S. (2012). The definition, assessment, and mitigation of state boredom within educational settings: A comprehensive review. *Educational Psychology Review, 24*(1), 89–111. https://doi.org/10.1007/s10648-011-9182-7
Voltaire. (1838). À madame Denis, nièce de l'auteur. La Vie de Paris et de Versailles (1748). In *Œuvres de Voltaire* (Vol. 34, pp. 168–174). Purrat Frères.
Voltaire. (2003). *Philosophical letters*. Dover Publications.
Voltaire. (2017). *Candide; or, the optimist*. Delphi Classics.
von Kleist, H. (2003). *Die Familie Schroffenstein*. Stroemfeld/Roter Stern.
Warren, H. C. (1934). *The dictionary of psychology*. Houghton Mifflin Company.
Watt, J. D., & Vodanovich, S. J. (1992a). Relationship between boredom proneness and impulsivity. *Psychological Reports, 70*, 688–690. https://doi.org/10.2466/pr0.1992.70.3.688
Watt, J. D., & Vodanovich, S. J. (1992b). An examination of race and gender differences in boredom proneness. *Journal of Social Behavior and Personality, 7*(1), 169–175.
Watt, J. D., & Vodanovich, S. J. (1999). Boredom proneness and psychosocial development. *Journal of Psychology, 133*(3), 303–314. https://doi.org/10.1080/00223989909599743
Zuckerman, M., Kolin, E. A., Price, L., & Zoob, I. (1964). Development of a sensation-seeking scale. *Journal of Consulting Psychology, 28*(6), 477–482.

Josefa Ros Velasco is a Marie Sklodowska-Curie Actions postdoctoral researcher at the Complutense University of Madrid, where she is the PI of the project "Pre-bored. Well-being and prevention of boredom in Spanish nursing homes" (2021–2024). Previously, she was a postdoctoral researcher for the Postdoctoral Research Fellowships at Harvard University for Distinguished Junior Scholars program (2017–2021). She is a specialist in Boredom Studies from a multidisciplinary perspective, and the founder and president of the International Society of Boredom Studies. She is the author of the book *The Disease of Boredom* (Princeton University Press, 2025), and editor of books such as *Suicide in Modern Literature* (Springer, 2021) or *Boredom Is in Your Mind* (Springer, 2019). For her academic achievements, she received distinctions such as the Spanish National Research Prize 2022 or the Harvard University Lincoln Book Prize 2019, among others. josros@ucm.es | @JosefaRosUCM | www.josefarosvelasco.com.

Part III
Hans Blumenberg's History and Philosophy of Technology

Technique, Technology, and Theory. Blumenbergian Perspectives in Philosophy of Culture

Christine Blaettler

Abstract Objects and concepts of technology were a lifelong theme in Blumenberg's work. In a broad sense, this comprises cultural techniques such as observation and description, rhetorical devices or individual helpful tools as well as the atomic bomb with its destructive power, and the entire field of modern technology. Blumenberg worked on the project of an 'intellectual history of technology' against a philosophical hostility towards technology that was widespread in the Federal Republic of Germany at his time, and considers technology in all its forms as a human cultural form. With this in mind, the article looks at Blumenberg's understanding of theory and works out its technical dimension. In contrast to the ancient understanding of *theoria* as contemplation, Blumenberg already identifies a technically shaped understanding of theory in the ancient science, and gains an 'epistemological model function' from astronomy: here the gaze is not content with contemplation, but follows a theoretical curiosity that makes use of technical instruments in addition to speculation and mathematics. Against a widespread juxtaposition of theory and practice, the article outlines Blumenberg's understanding of theory in its intervening dimension.

The writings of the 'technology aficionado' (B. Blumenberg, 2015) show an ongoing philosophical engagement with technology. On the one hand, Blumenberg research focuses on technique as craftsmanship, especially on the part of readings dedicated to the metaphorological works. Anselm Haverkamp, for example, locates "Blumenberg's project" in the "technique of rhetoric"; he reads the "Anthropological

---

A first version of this article was published in German under the title: "Technik der Theorie. Kulturphilosophische Perspektiven nach Blumenberg" (Blaettler, 2023a).

C. Blaettler (✉)
Christian-Albrechts-Universität zu Kiel, Kiel, Germany
e-mail: blaettler@philsem.uni-kiel.de

A. Fragio et al. (eds.), *Hans Blumenberg's History and Philosophy of Science*, Studies in History and Philosophy of Science 63,
https://doi.org/10.1007/978-3-031-78251-0_7

Approach to the Actuality of Rhetoric" as a "meta-rhetorical radicalization" (Haverkamp, 2001, p. 440). Rüdiger Campe (Campe, 2009) argues that Blumenberg's thematic interest in modern technology turns to rhetoric as *techne*: the question of technology shifts to the study of metaphors, technology thus becomes technique as rhetorical practice. In this sense, Campe defines modern technology as the "opening move" (Campe, 2009, p. 238) of Blumenberg's metaphorology, even if this was initially merely a "side branch" (Zill, 2017, p. 298) for Blumenberg.

The two posthumously published volumes entitled *Geistesgeschichte der Technik* and *Schriften zur Technik* focus on the man-made world of things and a modern understanding of technology (E. Müller, 2014). This focus is primarily linked to an argument of philosophical anthropology that ties technology back to the thesis of the human being as a deficient being: the human being creates its world of things on the basis of its imperfect physical endowment (O. Müller, 2005; Moxter, 2011). Blumenberg also explicitly refers to this thesis himself, but neither exclusively nor with an intention of defining it, quite the contrary. *Beschreibung des Menschen*, also published posthumously, brings together numerous specific definitions of the human species, not only taken from the philosophical tradition. Their compilation and the plurality they present means that each individual definition loses its distinction as decisive *differentia specifica*. Blumenberg draws attention to their character, which virtually parodies a definition, and aptly describes them as "definition essays" (Blumenberg, 2006, pp. 511–518): tentative possibilities and abbreviated descriptions of what could be understood as specifically human. Blumenberg advises testing these definition essays to see what they lack.

Blumenberg's project of an 'intellectual history of technology' is not articulated via the familiar and narrow anthropological understanding that explains technology from the human deficiency character. Instead, it starts with the aspect that understands technology as a historical phenomenon. This aspect "certainly includes the anthropological aspect of a deficient being, but transcends it insofar as technology, seen from this perspective, does not end up being an instrument for securing existence and satisfying elementary needs" (Blumenberg, 2015, p. 138). Blumenberg's point is rather that the human being "perceives its technicity and grasps it as the subject and signature of its self-interpretation and self-realization" (Blumenberg, 2015, p. 138). Technology is thus addressed poietically as a cultural form. The fact that the "intellectual history of technology" did not result in a monograph cannot be considered evidence of a "failure" (Goldstein, 2020, p. 133) of this project. Even if the later works are not part of a philosophy of technology in the narrower sense, technology continued to shape Blumenberg's thinking and proved to be a "life theme" (Zill, 2017, p. 291). Understood as a "question horizon", this project inspired the major monographs *Legitimacy of the Modern Age*, *Genesis of the Copernican World*, and *Work on Myth*, and it "also motivated" (Goslar and Voller, 2013, p. 424) the edition of Galileo's *Sidereus Nuncius*, for example. The following considerations attempt to explore Blumenberg's concept of theory using technology as a guideline and to outline cultural-philosophical perspectives on it.

1 The Sky and Its Knowledge

Based on the currently prevailing understanding of theory, it seems to be a kind of instruction manual, with its value depending on its successful applicability. Accordingly, theory should not only relate to practice, but should also prove itself in practice. In an academic context, theory should explain something, for example as a theory of culture, or it should guide action, as in applied ethics. Is theory seen here as the handmaiden of practice? On the other hand, we are witnessing a tendency to abandon theory altogether. Since the "practical turn" (Rheinberger, 2006, p. 17), theory has become increasingly less valued in scientific research. The epistemological revaluation of practical, namely experimental, research took place in the face of a theory of science focused on questions of justification (Schickore & Steinle, 2006), which in turn experienced a devaluation. The relationship between theory and practice remained asymmetrical: while practical-explorative experiments are welcome as knowledge-generating, theory-driven experiments (Steinle, 1998) are considered redundant at best. They would only serve to test a pre-existing theory and would not generate any new knowledge. Furthermore, theory is seen to be a detached and arbitrary construction, and this construction itself becomes absolutized as a subjective power of positing, which would make direct access to reality impossible and thus would miss it (Ferraris, 2012). Such an approach makes theory not only superfluous, but suspect. Talking about theory today, therefore, has its pitfalls. It may be more attractive to proclaim its weakness and even its end than to deal with how this suspicion of illegitimacy came about and what consequences it entails. Against the background of the current discourse going far beyond science and technology studies, some aspects of Blumenberg's version of a 'prehistory of theory' will be recalled here. Not least, this clearly indicates the systematic significance of the history of astronomy for Blumenberg's concept of theory and his approach in philosophy of culture.[1]

Thales of Miletus, who is considered the first philosopher in the European tradition, was a sky gazer. Yet, looking up at the sky is anything but a matter of course. Blumenberg repeatedly points out that it was only by walking upright that the human species was enabled to swap its everyday, practical, and vital close-up view for a distant one. Blumenberg (Blumenberg, 1966) identifies the starry sky, which has been effectively equated with the cosmos since antiquity, as the epitome of this distant vision. The view of the stars, which has been shaped by the history of human development, has moved the imagination, as we can see in myths and stories, utopias, and science fiction, in dreams, hopes, and fears up to this day. Blumenberg first identifies the "main source of belief in the gods" (Blumenberg, 1975, p. 29) in the view of the starry sky. He then draws attention to the exemplary function which the

---

[1] Alberto Fragio Gistau (2015) has also documented Blumenberg's interest in the history of astronomy and the astronomical research of his time on the basis of his *Nachlaß*, but has pursued a different path with regard to metaphorology and in close connection with Heidegger's existential ontology.

admiration of the heavens has for the "spectator idea" (Bumenberg, 1966, p. 116) with its implicit relationship between the human being and the cosmos. He correlates the "type of spectator" (Blumenberg, 1975, p. 512) with the ancient concept of reality insofar as it requires a static spectator. To this spectator, the phenomena do not simply appear, but they show themselves, the events present themselves and thus it is the world that acts. Accordingly, it is the sky that moves, not the people. Blumenberg uses the idea of the spectator to remind us that today's understanding of theory is virtually opposed to the ancient *theoria* as contemplation. This *theoria* was a matter of the leisure of a resting spectator, to whom the objects present themselves "shadow-free" (Blumenberg, 1975, p. 26).

The fact that this *theoria* was an ideal becomes clear from the theoretical primal scene introduced by Blumenberg. To Blumenberg, the sky-gazer Thales points out the "epistemological model function" (Blumenberg, 1966, p. 122) of astronomy: its objects seduce to contemplation, but at the same time elude idle vision. Thus, looking at the sky stimulates both contemplation and theoretical curiosity, because the naked human eye requires not only speculation and mathematics but also the reinforcement of tangible technical instruments. Blumenberg's observation of the sky thematizes a tension between the visible and the non-visible: it is hard to imagine a more distant object than the sky and its phenomena. At the same time, the sky promises evidence of the visible through the human sense of sight and its technical aids, be they polished glasses, simple telescopes and large astrophysical telescopes or computer-based virtual world telescopes. The history of astronomy through to space technology makes it clear how researchers always have to deal with the limits of the visible. They shift these limits with instruments underpinning this with visual representations. Such a recent shift in the boundary between the visible and invisible was celebrated as an event in 2019 with the Event Horizon Telescope. This giant telescope simulated from several analogue telescopes, which has a diameter comparable to that of the Earth, was used to visualize an image of the black hole at the center of the galaxy M 87 (Lutz, 2019). This event too is an example of Galileo's experience with the telescope as described by Blumenberg. As an event it provides a contemporary example of how visualization and mathematical construction conflict with each other, drive research and aim for harmonization, as the Event Horizon Telescope also generated its epoch-making image of the black hole with the help of gigantic computing power. Nevertheless, this virtual world telescope also denies an "optical totality" (Blumenberg, 1975, p. 717) of the cosmos and points back to partial perspectives, even in human-technical configurations.

Astronomy is considered an observational science that cannot experiment like other sciences, and even today's astrophysics cannot test its hypotheses empirically and experimentally. However, these areas of research make it clear that the contrast between active experimentation and passive observation does not apply. Observation cannot be equated with the ancient vision. There is no need to refer to concepts such as the theory-ladenness of observation, the underdetermination of a theory by data or the situatedness of the research arrangement (Bonk, 2008; Hanson, 1958; Haraway, 1988). But even experimentation cannot be understood solely in terms of human activity with the 'experimental turn' (Ackermann, 1998), if one takes into account that "[e]xperimentation has a life on its own" (Hacking, 1983, p. 149f.). The intervening character of research activity emphasized by Ian Hacking and attributed to experimentation must also be considered in astronomical observation (Cleland, 2002). In contrast to static observation, the former is methodical observation in which both intellectual and material instruments are used. In Galileo's case, this included cosmological theories, astronomical calculations, telescopes, descriptions, and drawings.

The practical or experimental turn has also found its disciplinary expression in astrophysics. In *Cosmic Discovery*, Martin Harwit (Harwit, 2019) noted with regard to the development of his discipline that important discoveries often did not result from theories but followed the use of a new observation instrument. However, in view of the practical emphasis in scientific research one should remember in this context that the development of instruments involves theoretical work, and that theories are embodied in the technical instruments of experimental science (Bachelard, 1971). It should also be borne in mind that in the history of astronomy it does not only hold true for Copernicus that calculations preceded discoveries which were then visually confirmed by new instruments. When today's astronomy and astrophysics are characterized as "very strongly driven by observation and technology" (Anderl, 2017, p. 82), it is primarily a matter of having the first possible access to the data provided by the latest technology. Indeed, there are complaints about an associated lack of methodological reflection and theoretical work, which would lead to "entirely baseless speculations" (Hossenfelder, 2020). Today's academic cosmology is considered an empirical science insofar as it is based on data that extends human perception far into the macrocosm (Anderl, 2017). However, how the data is linked to cosmic phenomena depends on their astrophysical classification, for which there are several justifiable options. Because of this plurality of systems, it is not possible to assert a cosmic order that is independent of human beings (Ruphy, 2016).

To Blumenberg the astronomer Galileo was the model of a phenomenology that considers visibility together with intentionality, perspectivity and technology, especially with regard to the tension between looking and doing. This phenomenology takes the conditions of the respective observation situation into account and, like Hacking, understands observation itself as a "skill by training and practice" (Hacking, 1983, p. 168). Blumenberg's Galileo presents its own epistemological-methodological "Galileo affair" (Stengers, 1989), whereby the approach developed is still worth considering: on the one hand, neither the loss of visibility would be seen as disenchantment, nor would appearance lead to false consciousness. On the other hand, the pleasure of vision (Blumenberg, 1975) is not linked to a desire for immediacy and the expectation of self-evidence, but rather to theoretical-technical mediality.

Since its philosophical beginnings, theory has operated via instruments, whether in the form of basic cultural techniques such as showing, writing, and calculating, or in the form of the artistry of methodical observation and mathematics. This includes "means of thinking" (Goldstein, 2020, p. 173), such as argument and causal relation or technical objects, such as optical instruments and space technology. In contrast to resting contemplation, a theoretical curiosity questions the prevailing order of the cosmos, which is supported by appearance and images, ideas, and dogmas. To an understanding of *theoria* as an evident view, such an intervention must appear as presumptuous and the theorist as a 'perpetrator.' Blumenberg's formula of the "theorist as perpetrator" (Blumenberg, 1975, p. 24, pp. 310–340) highlights a specific ambiguity: on the one hand, it designates a will to knowledge that becomes active and thus intervenes in an existing order. On the other hand, it addresses a judgment that condemns theoretical intervention in an existing order as human hubris.

In the tense constellation of ancient Greek *theoria* and ancient astronomy, the pre-Socratic philosopher Thales can already be regarded as a 'theorist perpetrator.' Today, this condemnation is echoed in the sweeping opposition to construction as represented by new realisms, speculative materialisms and object-oriented ontologies. These approaches claim to finally turn back to reality itself, without a theoretically mediating subject. While they make Kant an iconoclastic destroyer and constructivist caricature (Latour, 1988), Blumenberg portrays Kant as a theorist in the tradition of modern astronomy and uses him to unfold the ambiguity of the perpetrator. At the same time, Thales and Kant make it clear why Blumenberg is talking about theory and not philosophy. Before its historical differentiation, astronomy was part of the philosophical curriculum, as can be seen, for example, in the educational program in Plato's *Politeia* or in the mathematical quadrivium of the medieval faculty of arts. The "protophilosopher" (Blumenberg, 1987a, p. 11ff.) Thales was also an astronomer, and Kant had presented a *Universal Natural History and Theory of the Heavens* before his famous three *Critiques*. Kant's celestial treatise, which attempted to explain the origin of the planetary system without recourse to a divine ordering power, but was compatible with such a power, has gone down in the history of astronomy as Kant-Laplace's theory. This historical background not only allows theory to be understood as a broader field than today's philosophy, it also relates to both science and philosophy and evokes their common history of origin.

2 Theoretical Intervening

The active character of theory emphasized by Blumenberg can be understood in the sense of Hacking's intervening, thus highlighting its explorative and technical side. Observation and experimentation are no longer opposites insofar as the intervening aspect is taken into account in observation. 'Technical' is not defined here as the act of purpose-oriented production, nor does it reduce theory to a means-end thinking or instructions (see Recki, 2011). Instead, it marks a very fundamental point: something does not remain as it is, and this also means that something does not remain as self-evident as before. This emphasis on intervention has two advantages: on the one hand, it allows the renewed dispute between the schools of constructivism and realism to be undermined, as the concept derived from research practice resists a detached understanding of construction and does not fit easily into a constructivist theoretical tradition. On the other hand, intervention has the advantage of not immediately assigning a concept of action to an activity, whether this is strong in the sense of a subject with the power to act, or weak in the sense of an agency within the frame of the actor-network theory. In both instances this might be an overhasty attribution of Blumenberg's intervention to practical philosophy, namely ethics, pragmatism, or an object-oriented ontology. Intervening, on the other hand, allows the explorative and technical character of theory itself to become the subject. Along the lines of technique and technology, things come into view for philosophy that are manufactured and produce effects, that have a historical index, occur in a certain constellation, bring their own questions, and provoke others. This philosophy does not take its starting point from timeless entities, methods, or definitions, but from practices and artifacts, and considers their respective conditions and contingencies. Blumenberg's writings draw attention to the fact that describing, imagining, questioning, or methodically proceeding are both skills and activities, and that theory can be understood as technical insofar as it is produced on the one hand and can intervene on the other. The model of astronomy manifests a tension between theory and technology, which Blumenberg turns into an intra-theoretical matter with different historical manifestations—this can also be understood as an unfolding of the 'intellectual history of technology.'

The question now arises as to how Blumenberg himself intervened theoretically, be it in the history of philosophy and its reception or in the debates of his time. Even if this question can only be addressed in part here, the following examples from the field of a philosophical reflection on technology provide an insight. In so doing, Blumenberg's approach requires particular attention. It can be observed how he changes the understanding of technology by means of various descriptions, and above all hostility towards technology, that was widespread in the philosophy of his time (Zill, 2002). This includes activating pre-modern references to *techne* and *ars* and relating them to techniques and technologies of his time. In this way, the current understanding is extended to the earlier one, without, however, asserting an existential or valuing difference between modern and pre-modern technology. The first example of his approach is a reference to Marx in his project of an intellectual history of technology. At first glance, this project name seems to promise nothing less than an anti-Marx. But Blumenberg's passing mention of a footnote in Marx's *Capital* becomes relevant for his own project. Through Marx and the associated dismissal of Ernst Kapp, Blumenberg (2015) identifies his own project as both historical and critical (Blaettler, 2019a). In the aforementioned long footnote, Marx not only envisages the project of a "critical history of technology" (Marx, 1956–1990, 23, p. 392), he also marks a difference to an "abstract natural-scientific materialism that excludes the historical process" (ibid., p. 393). The "astonishing factual coincidence" (E. Müller, 2014, p. 334) with a Marxian concern observed elsewhere is also expressed here quite undogmatically.

Nevertheless, Blumenberg's work was recently categorized as "conservative cultural criticism" and its author described as the "Walter Benjamin of conservatives" (Habel, 2019). This bon mot is symptomatic, as it testifies to the well-known misunderstanding that still assumes agreement with thinkers such as Martin Heidegger,

Ernst Jünger, or Carl Schmitt in Blumenberg's reception. However, Blumenberg is considered conservative above all with regard to the bourgeois educational canon and his philological erudition in ancient literature, too. Today it is easy to declare his writings eurocentric, androcentric, or elitist—although this is true of most research in all the more traditional disciplines. If one considers Blumenberg's approach and the way he proceeds, things look different. The philosophical canon, insofar as he took it up alongside many forgotten sources, served him as both material and instrument, and he researched settings, reassignments, and counter-assignments, and he did rereading and rewriting the canon (Blumenberg, 1987a, 1989). On the one hand, this expresses neither confirmation nor justification of the existing, and on the other, these procedures continue to invite theoretical debate.

Manfred Sommer has emphasized Blumenberg's conspicuously non-explicit, indirect, roundabout, and retrospective engagement with Heidegger and pointed out "the extent to which Blumenberg philosophizes 'against' Heidegger as a whole" (Sommer, 2013, p. 417; see Blaettler, 2023c). A second example of Blumenberg's approach is an indirect discussion with Heidegger's writings on technology. To this day, Heidegger's *The Question Concerning Technology* is widely regarded across disciplines as the first reference for a philosophical approach to technology in general. On the one hand, his reconsideration of the ancient understanding of *techne* is central to this, although he was neither the only nor the first one to introduce it, and on the other, his pessimistic outline of modern technology as a frame and fate (Blaettler, 2016; Morat, 2007). Blumenberg refers to both aspects, and almost without direct reference he takes up Heidegger's claims and parries them with a different twist, especially where it is not immediately obvious. Blumenberg's mytho-theoretical commitment and his elaborations of the variations and ways of the Prometheus mythologem, can thus be understood in *Arbeit am Mythos* as an objection to the 'modern archaics' (Voller, 2014; see also Blaettler, 2024b). In the German post-war period, conservative thinkers defined Prometheus, the mythical bringer of culture and "proto-technician" (Voller, 2014, p. 278), as an outlaw, identified him with the hubris of 'modern man', and justified his divine punishment once and for all by referring back to a myth that was originally set in stone. In view of such a direct transfer to the discourse on technology, Blumenberg's metaphorological considerations make it clear not least how Heidegger's writings, in their direct literal and etymological paths, bear witness to forgetfulness of metaphors (Blumenberg, 2014).

Even in his earliest writings, Blumenberg (2015) distances himself from the widespread fatalization and demonization of technology as an autonomous agency, which can also be found in Heidegger, and instead links technology with human degrees of freedom (see Zill, 2017). Blumenberg neither glorifies pre-modern technology nor does he reduce the human being to the modern *Homo faber* who exploits nature. The technology problematized in his present served Blumenberg as a starting point for giving this technology an 'intellectual history': on the one hand, questions about nature versus technology, about technical progress and processes of mechanization are given a historically deep dimension through *techne* and *ars* as artistry, cultural technique, practical and professional knowledge. On the other hand, Blumenberg crosses the ontological and normative definition of pre-modern and modern technology through the historical-dynamic relationship between nature and technology as well as through stages of imitation and recreation of nature. For Blumenberg, technology is not applied science either, but a cultural form in its own right, even if the two forms interact. Blumenberg's writings on technology and science implicitly articulate an objection to the suspicion of illegitimacy of modernity raised by his conservative contemporaries who were involved in National Socialism. In this sense, Blumenberg's technology project not only proves to be a phenomenological alternative to Heidegger's question of technology, it also remains a philosophical commitment worth considering in the face of hostility to technology coupled with hostility to modernity, even in the Anthropocene discourse.

Apart from the fact that Heidegger offers a "Dorado for the fun to parody" (Blumenberg, 1997, p. 92), Blumenberg's method appears less as parody (Haverkamp, 2009) or mimesis than as a demonstration of an experiment, and not only in his discussion of Heidegger: a claim of validity is taken as a point of departure in order to test what degrees of freedom, other paths and differences are possible in an existing discursive field of forces, or at least could be possible, and it is all about the difference in its entirety. Blumenberg's analyses of historical examples sometimes contain passages that sound like self-descriptions. Two examples from the history of astronomy will illustrate this. The first passage concerns Copernicus' relationship with Ptolemy, with whose conclusions Copernicus largely disagreed. Blumenberg draws attention to a "respectful understatement" with which Copernicus addresses Ptolemy less as an enduring authority than as the "founder of possibilities" (Blumenberg, 1986, p. 118). He emphasizes that this is the "expression of the effort" to "counter the 'scholastic' procedure of submission to an authority that can only be commented on with the rational self-preservation of the conditions" of "theory as *humanly possible*" (Blumenberg, 1986, p. 124).

The second passage refers to Galileo. His *Dialogue Concerning the Two Chief World Systems* is repeatedly cited as evidence of a cave-in to the ecclesiastical authorities. Blumenberg also reads the *Dialogue* with an eye to its political-tactical function after the Index Decree of 1616, which prescribed the hypothetical form for all Copernican considerations. However, he accentuates Galileo's handling of the theological reservation of sovereignty by turning his attention from the imposed end of the dialogue to the oldest component of the Tidal Treatise. In this, Galileo is factually mistaken but, according to Blumenberg, rhetorically brilliant. He identifies Galileo's skill in incorporating the omnipotence clause into his own train of thought as an "argumentative foreign body" (Blumenberg, 1975, p. 493f.; see also Blumenberg, 1965) and encapsulating it within it. In doing so, Galileo turns the original intention of the omnipotence clause into its opposite and in favor of his own scientific concerns. Both passages document Blumenberg's attention to forms of criticism that are formulated neither directly nor negatively or explicitly polemically, but which are critical in substance. A claim to validity is taken into account, but theoretically 'disempowered,' so to speak, without explicitly rejecting or negating it. This criticism is articulated indirectly and can also be used as a method against arguments and theories. Once these have entered the methodological toolbox, they are automatically applied and run through, so to speak. They have become commonplace and run the risk of detaching themselves from their concrete factual reference, of changing or even losing their initial intention—of becoming absolute. Paying attention to such dynamics makes it a real task to question self-proclaimed criticism as to whether and how this criticism is actually carried out.

To what extent is Blumenberg's own very roundabout critique a method against the "barbarism" (Blumenberg, 1987b, p. 137) of the straight path that strikes directly? His understanding of method as a detour also applies here. Motivated by the matter to be examined, it manifests itself in the fact that a matter is pursued (Blaettler, 2019b; Blumenberg, 2015; Zill, 2015). In critical examination, the detour proves to be an explorative procedure, insofar as it does not confirm or deny something that is already known, but rather searches for new possibilities and paths. The ambiguity of the verb *verfahren* ('to proceed' and 'to lose one's way,' 'to detour') illustrates this aptly: it is about proceeding methodically and deviating from the direct path. Even if detours can be very skillful, the focus here is neither on *techne* nor on the definition of metakinetic "time modulators" (Campe, 2009, p. 291), according to which rhetoric slows down and technology speeds up. Rather, in contrast to a direct means-end relationship and an understanding of technology as the "epitome of means" (Hubig, 2006, p. 107), the indirect and circuitous route thematizes the intervening indirectness of technology in all its forms. Finally, an example will be used to explain how Blumenberg's reformulation of philosophy as a philosophy of culture is characterized by such a concept of technology, which makes the friction between vision and intervening identified in the theoretical primal scene productive.

3 Neither Worldview Nor Iconoclasm

Currently, grand new narratives of progress and decline are emerging and gaining astonishing popularity that are in no ways inferior to the old ones. This finds its expression, for example, in apocalyptic imaginings or in the cosmological and world-historical picture made via the epochal category of the Anthropocene and the Gaia hypothesis. It is striking how, following the often quoted "end of the grand narratives," the anthropocene discourse invokes "huge time horizons of geological transformations" (Horn, 2017, p. 11) that make Hegel's philosophy of history—for many still today the epitome of totality and even suspected of totalitarianism (Ricœur, 1985)—seem almost cute in comparison. While an atmospheric chemist launched the word simply as a suggestion to refer to an epochal concept, grand narratives have emerged since then. The Anthropocene thesis sees "humanity as the driving force behind the planetary transformations of our planet" (Scherer and Klingan, 2013, p. 3) and combines this with ecological concerns that identify the human factor as the cause of all evil. By using the word 'human' to determine the naming of a geological epoch, the idea of a significantly increased, indeed immense power of 'the human being' is expressed. Its power of disposal over the entire earth is even addressed as a total one, if there is to be "no more nature *beyond* human control" (Horn, 2017, p. 8). The understanding of technology that emerges in climate debates is based on the idea of a consciously acting human being who primarily controls 'nature' through technology. However, instead of problematizing this one-sided idea of 'the human being,' the species is generally incriminated and the familiar critique of modernity formulated about technology and science is repeated, which does not so much analyze the issue as approach the problem normatively. Bruno Latour's *Gaia Lectures* (2017), for example, reveal an approach of this kind in that they not only make a difference between good and evil, but also use the decisive difference of political theology between friend and foe to identify guilty parties and evoke an ecological world war (see Werber, 2015).

Such grand narratives can be described as new cosmologies insofar as they metaphorize scientific concepts into a world view on the one hand (Blumenberg, 1960), and reactivate an old idea of philosophy as an explanation of the world on a cosmic scale on the other. It is hardly surprising that these new cosmologies are based on strong historical-philosophical assumptions. These assumptions are either committed to a technophile optimism about progress, but more often to a culture-critical pessimism about doom. In the case of the Anthropocene, it can be observed how a scientific model for a "total conception of empirical reality" was mobilized into a world model, which in turn was promoted with a "normative epitome of reality" (Blumenberg, 2015, p. 128) to a comprehensively meaningful world view. The category of the Anthropocene also raises major questions that are by no means unjustified in their concern as questions of meaning, but whose answers are problematic, be it with regard to their normative claim to validity or their practically enforced validity. Based on the experience that worldviews provide "almost inexhaustible legitimation for actions of every kind and quality" (Blumenberg, 1961, p. 138), Blumenberg called for a determined refusal of worldviews in philosophy, i.e., not to provide "answers to the final and most comprehensive questions" (Blumenberg, 1961, p. 140). It is not the big questions that are the problem, but the big answers. The call for a refusal of worldviews in philosophy should not be understood to mean that it is a matter of clearing up misconceptions. Nor does this refusal lead to a philosophical iconoclasm that simply discredits images of the world, myths and metaphors, ideals and idols. Ideas that extend to "total horizons" (Blumenberg, 2014, p. 90) express a historical "space of experience" and "horizon of expectation" (Koselleck, 1989, pp. 349–375) and, as cultural phenomena, can become an object of description, analysis and further philosophical debate.

In Blumenberg's case, this becomes particularly clear in view of the anti-modern and pessimistic cultural criticism that is often articulated through hostility towards technology, which no longer imagines the future as an open horizon, but only as a disastrous downfall. After the introduction, the volume *Ein mögliches Selbstverständnis* immediately brings together short texts under the title "Endzeittümelei," which no longer trace the topic in the historical examination, but in Blumenberg's epoch. The apocalypse skepticism expressed in these texts is not only remarkable in view of the apocalypse blindness claimed by Günther Anders (Anders, 1994), it is also relevant in view of the apocalyptic desire that has emerged in the ecological debate. How are the present and the future imagined here? And to what extent does the fear of the end, which is invoked as necessary, have a community-building effect, who is included and excluded, and to what extent is the present devalued not only in relation to the past, but also to the future? Blumenberg already identified doomsday agonizing among the conservative critics of modernity of his time. They stoked fears of doom and counted modernity as a whole, especially modern science and technology, as the cause and promoter of this decline.

In contrast to the "absolute claim" of apocalyptic narratives registered by Blumenberg (Blumenberg, 1997, p. 33), his philosophical insistence on contingency takes a different path. Here, contingency is not to be understood as simply transient, random, or arbitrary, but rather modally as something that is neither necessary nor impossible. Blumenberg's work shows how theories, scientific models and world views that have actually prevailed historically can be explored, described and questioned with regard to their respective conditions of possibility and histories of formation. "Historical knowledge of the power of worldviews" (Blumenberg, 1961, p. 135) in no way negates their real effectiveness. However, it disempowers them at the theoretical level and creates a basis for intervening theoretically instead of merely justifying their factual validity in theory. Contingency also becomes relevant for this philosophy precisely to the extent that it "dismantles the self-evident" (Blumenberg, 2001, p. 414) and testifies to a partial, provisional, and plural understanding of knowledge (Longino, 2001). This strategy is clearly manifested not only against fundamentalisms and world views, but also against theoretical absolutisms.

Philosophy in the name of culture is by no means self-evident and finds itself challenged anew. The prevailing understanding of culture today leads, at best, to misunderstandings that arise over the same word without taking conceptual differences into account. An 'ethnicization of politics' (Salzborn, 2005) attached to the word 'culture' as well as identitarian and biologistic naturalizations of culture are diametrically opposed to philosophy of culture as an enterprise in the tradition of the first cultural science (Weigel, 2021; see also Blaettler, 2023b, 2024a). Such a philosophy not only makes culture its object, but it is also a philosophy that reflects on itself culturally and addresses its own premises, methods and objects as entities that are themselves presuppositional and contingent. It does not define these entities in definitive, normative, or essentialist terms as supratemporal and general, and yet does not regard them as arbitrary and interchangeable. A philosophy committed to conditionality and contingency is currently facing accusations of weak thinking. It is found to be out of touch with reality, blamed for alternative facts and their vociferous political representatives and dismissed as unsuitable by a new self-proclaimed strong thinking (Ferraris, 2012). As much as objections to detached constructivism are justified, the question arises as to whether the new strong thinkers are not once again allowing themselves to be seduced into feeling "in league with the reason of history" (Blumenberg, 1961, p. 138) and confirming in theory what is already the case of the 'history of the winners'? And is not the strength of thinking grounded precisely in the fact that, in addition to the beautiful aspects, it also confronts an undesirable and uncomfortable, indeed a hard and ugly reality, but does not legitimize it with agonizing about doomsday scenarios, but intervenes theoretically?

References

Ackermann, R. (1998). The new experimentalism. *British Journal for the Philosophy of Science, 40*(2), 185–190.

Anderl, S. (2017). *Das Universum und ich. Philosophie der Astrophysik*. Hanser.

Anders, G. (1994). *Die Antiquiertheit des Menschen 1. Über die Seele im Zeitalter der zweiten industriellen Revolution*. C. H. Beck.

Bachelard, G. (1971). Le matérialisme technique. In *Épistémologie. Textes choisis* (pp. 136–157). PUF.

Blaettler, C. (2016). Heideggers Technikbegriff und der verstellte Blick auf die Ökonomie. *Navigationen, 2*, 65–81.

Blaettler, C. (2019a). Natürliche Technologie und technische Differenz. Ernst Kapp und die Frage der Biologie für den Kulturapparat, gelesen mit Kant und Blumenberg (pp. 149–174). *Jahrbuch Technikphilosophie*.

Blaettler, C. (2019b). Philosophie, als Verfahren. In B. Kaiser & H. Schmiedl-Neuburg (Eds.), *Philosophie und Literatur* (pp. 29–48). Bautz/Traugott.

Blaettler, C. (2023a). Technik der Theorie. Kulturphilosophische Perspektiven nach Blumenberg. *Allgemeine Zeitschrift für Philosophie, 2*, 195–216.

Blaettler, C. (2023b). Wider den Kulturpessimismus. Vier Thesen zu Philosophie im Namen von Kultur. *Zeitschrift für Kulturphilosophie, 2*, 35–42. https://doi.org/10.28937/978-3-7873-4506-9

Blaettler, C. (2023c). Sorge und Welt: Blumenberg versus Heidegger. *Archiv für Begriffsgeschichte, 1*, H. 65/2.

Blaettler, C. (2024a). Kultur und Kritik. *Y – Zeitschrift für atopisches Denken. Philosophie – Psychoanalyse –Kulturwissenschaften*, 2024, *4*(3). https://www.ypsilon-psychoanalyse.de/tribuene/167-kultur-und-kritik

Blaettler, C. (2024b). Fear and freedom in technology. On prometheus as an ambiguous figure of technology in philosophy. In J. Allegrante, U. Hoinkes, M. Schapira, K. Struve (Eds.), *Anxiety culture: The new global state of human affairs* (pp. 292–304). Johns Hopkins University Press.

Blumenberg, B. (2015, July 13). Eine Frage der Belichtung. *Neue Zürcher Zeitung*.

Blumenberg, H. (1960). Paradigmen zu einer Metaphorologie. *Archiv für Begriffsgeschichte, 6*, 7–142.

Blumenberg, H. (1961). Die Bedeutung der Philosophie für unsere Zukunft. *Europa-Gespräch 1961: Die voraussehbare Zukunft*, 127–140.

Blumenberg, H. (1965). Das Fernrohr und die Ohnmacht der Wahrheit. In G. Galilei (Ed.), *Sidereus Nuncius. Nachricht von neuen Sternen* (pp. 7–75). Suhrkamp.

Blumenberg, H. (1966). Contemplator Caeli. In D. Gerhardt (Ed.), *Orbis Scriptus. Dimitrij Tschizewskij zum 70. Geburtstag* (pp. 113–124). Fink.

Blumenberg, H. (1975). *Die Genesis der kopernikanischen Welt*. Suhrkamp.

Blumenberg, H. (1986). *Lebenszeit und Weltzeit*. Suhrkamp.

Blumenberg, H. (1987a). *Das Lachen der Thrakerin. Eine Urgeschichte der Theorie*. Suhrkamp.

Blumenberg, H. (1987b). *Die Sorge geht über den Fluss*. Suhrkamp.

Blumenberg, H. (1989). *Höhlenausgänge*. Suhrkamp.

Blumenberg, H. (1997). *Ein mögliches Selbstverständnis*. Suhrkamp.

Blumenberg, H. (2001). Anthropologische Annäherung an die Aktualität der Rhetorik. In *Ästhetische und metaphorologische Schriften* (pp. 406–431). Suhrkamp.

Blumenberg, H. (2006). *Beschreibung des Menschen*. Suhrkamp.

Blumenberg, H. (2014). Ausblick auf eine Theorie der Unbegrifflichkeit. In *Schiffbruch mit Zuschauer: Paradigma einer Daseinsmetapher (1979/1997)* (pp. 85–106). Suhrkamp.

Blumenberg, H. (2015). *Schriften zur Technik*. Suhrkamp.

Bonk, T. (2008). *Underdetermination. An Essay on Evidence and the Limits of Natural Knowledge*. Springer.

Campe, R. (2009). Von der Theorie der Technik zur Technik der Metapher. Blumenbergs systematische Eröffnung. In A. Haverkamp & D. Mende (Eds.), *Metaphorologie. Zur Praxis von Theorie* (pp. 283–315). Suhrkamp.
Cleland, C. E. (2002). Methodological and epistemic differences between historical science and experimental science. *Philosophy of Science, 69*, 474–496.
Ferraris, M. (2012). *Manifesto of new realism*. State University of New York Press.
Fragio Gistau, A. (2015). *Paradigms for Metaphorology of the Cosmos. Hans Blumenberg and the Contemporary Metaphors of the Universe*. Aracne.
Goldstein, J. (2020). *Hans Blumenberg. Ein philosophisches Portrait*. Matthes & Seitz.
Goslar, T.-F., & Voller, C. (2013). Geistesgeschichte der Technik als 'Kritik der Fortschrittskritik'. *Zeitschrift für Kulturphilosophie, 2*, 423–428.
Habel, S. (2019, January 1). *Die Eskalation der Zeit. Hans Blumenberg als konservativer Diagnostiker der Moderne*. Geschichte der Gegenwart. https://geschichtedergegenwart.ch/die-eskalation-der-zeit-hans-blumenberg-als-konservativer-diagnostiker-der-moderne/
Hacking, I. (1983). Representing and intervening. In *Introductory topics in the philosophy of natural science*. Cambridge University Press.
Hanson, N. R. (1958). *Patterns of discovery*. Cambridge University Press.
Haraway, D. (1988). Situated Knowledges. The science question in feminism and the privilege of partial perspective. *Feminist Studies, 14*(3), 575–599.
Harwit, M. (2019). *Cosmic discovery. The search, scope, and heritage of astronomy*. Cambridge University Press.
Haverkamp, A. (2001). Die Technik der Rhetorik. Blumenbergs Projekt. In H. Blumenberg (Ed.), *Ästhetische und metaphorologische Schriften* (pp. 435–454). Suhrkamp.
Haverkamp, A. (2009). Das Skandalon der Metaphorologie. Prolegomena eines Kommentars. In A. Haverkamp & D. Mende (Eds.), *Metaphorologie. Zur Praxis von Theorie* (pp. 33–61). Suhrkamp.
Horn, E. (2017). Jenseits der Kindeskinder. Nachhaltigkeit im Anthropozän. *Merkur, 71*(814), 5–17.
Hossenfelder, S. (2020, January 8). Why the foundations of physics have not progressed for 40 years. *iai News, 84*. https://iai.tv/articles/why-physics-has-made-no-progress-in-50-years-auid-1292
Hubig, C. (2006). *Die Kunst des Möglichen I: Grundlinien einer dialektischen Philosophie der Technik, Bd. 1: Technikphilosophie als Reflexion der Medialität*. Walter de Gruyter.
Koselleck, R. (1989). *Vergangene Zukunft. Zur Semantik geschichtlicher Zeiten*. Suhrkamp.
Latour, B. (1988). The politics of explanation: An alternative. In S. Woolgar (Ed.), *Knowledge and reflexivity. New frontiers in the sociology of knowledge* (pp. 155–176). SAGE.
Latour, B. (2017). *Facing Gaia: Eight lectures on the new climatic regime*. Cambridge University Press.
Longino, H. E. (2001). *The fate of knowledge*. Princeton University Press.
Lutz, O. (2019, April 19). How scientists captured the first image of a Black Hole. *JPL NASA*. https://www.jpl.nasa.gov/edu/news/2019/4/19/how-scientists-captured-the-first-image-of-a-black-hole/
Marx, K., & Engels, F. (1956–1990). *Werke (MEW)*. Institut für Marxismus-Leninismus beim ZK der SED.
Morat, D. (2007). *Von der Tat zur Gelassenheit. Konservatives Denken bei Martin Heidegger, Ernst Jünger und Friedrich Georg Jünger 1920–1960*. Wallstein.
Moxter, M. (Ed.). (2011). *Erinnerungen an das Humane: Beiträge zur phänomenologischen Anthropologie Hans Blumenbergs*. Springer.
Müller, O. (2005). *Sorge um die Vernunft. Hans Blumenbergs phänomenologische Anthropologie*.
Müller, E. (2014). Technik. In R. Buch & D. Weidner (Eds.), *Blumenberg lesen. Ein Glossar* (pp. 323–336). Suhrkamp.
Recki, B. (2011). Auch eine Rehabilitierung der instrumentellen Vernunft. Blumenberg über Technik und die kulturelle Natur des Menschen. In M. Moxter (Ed.), *Erinnerung an das Humane. Beiträge zur phänomenologischen Anthropologie Hans Blumenbergs* (pp. 39–61). Springer.

Rheinberger, H.-J. (2006). *Epistemologie des Konkreten*. Suhrkamp.
Ricœur, P. (1985). *Temps et récit, vol. 3: Le temps raconté*. Editions Seuil.
Ruphy, S. (2016). *Scientific pluralism reconsidered. A new approach to the (dis)unity of science*. University of Pittsburgh Press.
Salzborn, S. (2005). *Ethnisierung der Politik. Theorie und Geschichte des Volksgruppenrechts in Europa*. Campus Verlag.
Scherer, B. M., & Klingan, K. (2013). Einführung. In *Das Anthropozän-Projekt. Eine Eröffnung* (pp. 2–7). HKW.
Schickore, J., & Steinle, F. (Eds.). (2006). *Revisiting discovery and justification. Historical and philosophical perspectives on the context distinction*. Springer.
Sommer, M. (2013). Hans Blumenberg. Abweisung auf Umwegen. In D. Thomä (Ed.), *Heidegger Handbuch. Leben – Werk – Würdigung* (pp. 417–420). Metzler.
Steinle, F. (1998). Exploratives vs. theoriebestimmtes Experimentieren: Ampères erste Arbeiten zum Elektromagnetismus. In M. Heidelberger & F. Steinle (Eds.), *Experimental Essays – Versuche zum Experiment* (pp. 272–297). Nomos Verlag.
Stengers, I. (1989). Les affaires Galilée. In M. Serres (Ed.), *Éléments d'histoire des sciences* (pp. 337–375). Bordas Editions.
Voller, C. (2014). Wider die 'Mode heutiger Archaik': Konzeptionen von Präsenz und Repräsentation im Mythosdiskurs der Nachkriegszeit. In B. Gebert & U. Mayer (Eds.), *Zwischen Präsenz und Repräsentation. Formen und Funktionen des Mythos in theoretischen und literarischen Diskursen* (pp. 226–257). Walter de Gruyter.
Weigel, S. (2021). *Intellectual potential from the other side of Europe's colonial and nationalist past. Cultural science/Kulturwissenschaft around 1900 and its relevance for cultural relations*. SSOAR.
Werber, N. (2015). Gaias Geopolitik. *Merkur, 69*(5), 59–67.
Zill, R. (2002). Im Wendekreis des Sputnik. Technikdiskurse in der Bundesrepublik Deutschland der 50er Jahre. In I. Schneider & P. M. Spangenberg (Eds.), *Medienkultur der 50er Jahre. Diskursgeschichte der Medien nach 1945, Bd. 1* (pp. 25–49). Westdeutscher Verlag.
Zill, R. (2015). Auch eine Kritik der reinen Rationalität. Hans Blumenbergs Anti-Methodologie. In M. Heidgen, M. Koch, & C. Köhler (Eds.), *Permanentes Provisorium. Hans Blumenbergs Umwege* (pp. 53–74). Brill.
Zill, R. (2017). Von der Atommoral zum Zeitgewinn: Transformationen eines Lebensthemas. Hans Blumenbergs Projekt einer Geistesgeschichte der Technik. In A. Friedrich (Ed.), *Jahrbuch Technikphilosophie* (pp. 291–313). Nomos.

Christine Blaettler has been a philosophy professor at Kiel University since 2011. She studied philosophy and Slavic languages and literature in Berne, Fribourg and Eastern Europe, was a research fellow in Berlin (ZfL), Vienna (IWM), Weimar (IKKM), and at Stanford, and a visiting professor at the University of Vienna. She works on the cultural forms of science, technology and art. Among her recent publications are: "Fear and Freedom in Technology. On Prometheus as an Ambigous Figure of Technology in Philosophy," in *Anxiety Culture: The New Global State of Human Affairs* (Johns Hopkins University Press, 2024); "Kultur und Kritik," in *Y Zeitschrift für atopisches Denken. Philosophie - Psychoanalyse - Kulturwissenschaften* (2024); "Hannah Arendt und die Frage nach der Geschichte," Nachwort, in Hannah Arendt, *Macht und Gewalt* (Piper, 2024); "Sorge und Welt: Blumenberg versus Heidegger," in *Archiv für Begriffsgeschichte* (2023); *Theoretische Neugierde. Horizonte Hans Blumenbergs* (*Allgemeine Zeitschrift für Philosophie*, 2023); *Angst und Freiheit. Theoretische und künstlerische Positionen* (Würzburg: Königshausen & Neumann, 2023); *Der Gesandte. Alexandre Kojèves europäische Missionen* (Merve, 2022); *Benjamins Phantasmagorie. Wahrnehmung am Leitfaden der Technik* (Dejavu, 2021).

Nobody's Driving. Reality and Technology According to Hans Blumenberg

Gonzalo Forsberg

Abstract This article schematically reconstructs Hans Blumenberg's theses on technology by going through his extensive published and posthumous work with a fine-toothed comb. Blumenberg's canonical takes on the themes of metaphorology, anthropology, and phenomenology are interpreted here according to a particular hypothesis, namely, that most of the German philosopher's upshots about these subjects coalesce into the question of technology and its transformative aspect upon reality. In order to develop such proposition, (1) I argue how the question of technology remained in his thought, from its very beginning to his posthumous writings, as an enticing yet ever-changing epistemic paradigm. Likewise, (2) I ponder why such theses on technology stands out in the philosophical constellation of twentieth-century debates on this subject. Finally, (3) I address the importance on such a theme by conveying Blumenberg's thesis on technology with his multileveled concept of reality. (4) A conclusion is also given apropos the actuality of Blumenberg's thought among post-2020 philosophical debates.

1 Introduction

Amebix is considered one of the most important genre-defying bands of 1980s underground music. Their unique blend of heavy metal, punk, and post-punk still mesmerize old and new fans around the world. Amebix lyrics engage with advanced technology, dystopic political landscapes, and modern psychological syndromes—such as paranoia, anxiety, and all sort of gruesome mental conditions. Unlike most of their contemporaries, Amebix offered a spiritual viewpoint, which manifested sonically through ominous sounds embedded in sinister yet lucid reflections. Their second album—entitled *Monolith*—was released in 1987, just one year after the

G. Forsberg (✉)
Leiden University, Leiden, The Netherlands
e-mail: gonzaloforsberg@proton.me

A. Fragio et al. (eds.), *Hans Blumenberg's History and Philosophy of Science*, Studies in History and Philosophy of Science 63,
https://doi.org/10.1007/978-3-031-78251-0_8

Chernobyl disaster. Much like Andréi Tarkovsky's astonishing foreshadowing of this event in *The Sacrifice* (1987), *Monolith* also signaled the feeling of chaos and haplessness amid the uncontrollable forces of nuclear energy. In the second track of the album, 'Nobody's Driving,' the chorus sings:

We are swimming in the lunar sea
Drowning in insanity
Look to the shore you will see
Your leaders were lying
Nobody's driving!

In the first chapter of his posthumous *Sources, Streams, Icebergs* (2021), metaphors-tracker Hans Blumenberg warns—apropos Freud's use of the metaphor of the iceberg—'how one has to fully listen to metaphors when they are taken verbatim' ['was aufhorchen lässt, wie es immer dan geschieht, wenn die Metapher beim Wort genommen wird'] (p. 31). In this case, that an underground English music outfit relies on the metaphor of the ship—refurnished by Otto Neurath (Blumenberg, 1997a) in the twentieth century as a critique to Rudolph Carnap's longing for immaculate constructed sentences—demands utmost attention. If one follows Blumenberg's proviso regarding metaphors, it is something quite intriguing the variation on a theme carried out by a 1980s punk-metal cult band. The 'ship simile'—taken by Neurath as an ongoing work in progress of metaphysics and language over the sea—is now revealed at its full extent: sailors realize in awe-struck that their ship is equipped not only with 'imprecisions' but, moreover, that they are headed towards a burial at sea. The vessel was always fled under no command. An already dramatic metaphor, now meets its darkest version, musically crafted after an unseen historical singularity.

Nevertheless, why a long-life metaphor about shipwrecking is used as a fair depiction of the aftermath of a nuclear accident? In fact, what is a metaphor as such? It is quite baffling that even in his introduction to *Paradigms for a Metaphorology* (2020a), Blumenberg does not provide a definition of what a metaphor is. As Anselm Haverkamp has pointed out, the idea of a metaphor 'excels and exceeded all history in the usual sense' (2012, p. 40). If this is true—and one can rightly confirm such a statement by combing out Blumenberg's published and posthumous published material, let alone his archives at the *Nachlass* in Marbarch—, then metaphors are conceptually housed in history. Nonetheless, what happens when non-historical events suddenly unleash out of nowhere? While Blumenberg did not offer a crystal-clear distinction between metaphors and platitudes, not even the tritest saying could have ever captured the ungodly energy liberated at Chernobyl. Even the terminus technicus 'actio per distans' (Zill, 2020, p. 524) falls short at the moment of offering a visual guidance or at least a mental sequence regarding Chernobyl. In sum: what to make with metaphors-free events? Otherwise put, what happens when the link between reality and technology is regulated by blunt asymmetry?

In the following, I will address this subject throughout a roughly reconstruction of Hans Blumenberg's thesis on technology—or, more precisely, *technisierung*—and reality. I consider that the intellectual efforts of Blumenberg during the 1980s aimed at this complex entanglement of technology and reality. That his 'metaphorology' somehow progressively fades away in his last books (Haverkamp, 2012, p. 41), nods to the methodological problems that ongoing *technisierung* set for a major philosophical inquiry. More than a metaphor for nuclear disasters or imminent technological apocalypses, Blumenberg sought for a long-scale, fully-fledged account of phenomenology. Non-conceptuality shifts towards soon-to-be technology-aware concepts. Maximum disasters yield maximum reflections.

2 Technology

It is not a far-fetched thesis to state that Blumenberg's exhaustive investigations aimed at the question of reality. Although the reader can easily grasp this by merely reading one of his collection of essays, entitled *The Realities We Live in* (1981a), Blumenberg's outstanding investigations would only have been labeled on the off chance as contributions towards a philosophy of reality. On the contrary, the immense length and breadth of his erudition—starting by the massive *The Legitimacy of Modern Age*—along his knowledge of the history of philosophy, drew the attention of Blumenberg's contemporaries as an exotic blend of a postwar Theologian, neo-Heideggerianism, and a creative Husserlian philosopher at the same time. However, since his conference at Brussels in 1953—entitled 'Technic and Truth' (2015a, p. 42)—the problem of reality, even before he coined the concept of 'metaphorology,' was already latent by the late 1940s. Blumenberg's interest in day-to-day (*Alltäglicheres*) events was developed under the Axel Colly moniker (such was the name of his dog) through philosophical journalism in the early 1950s, and then later in his honing use of anecdotes as epistemic postcards of the lifetime—the 'smallest unit of life watched from the outside,' as Rüdiger Zill ponders (2020, p. 29). The fact that Blumenberg filled up since the early days his intellectual life with throbbing considerations on reality, suggest this subject as the problem of philosophy. Pseudo-Plato's statement on 'the object itself, the knowable and truly real being' (Burnyeat and Frede, 2015, p. 123) would endorse this clumsy assertion. In a posthumous volume recently published, Blumenberg conceptually deepens the question of *realität*, by considering it, first, as a 'contrast concept' ['Kontrastbegriff']—for what we experienced as reality is the intensity of the unreal—and, later, as an 'implicative predicate' ['implikatives Prädikat'] (2020a, p. 39). One could say: reality comes after having received the unchosen visite of loss.

Reality unfolds itself through unrealistic episodes. One questions what is real – what is *really* real—at the very moment when reality as such present askew, unfamiliar features. Modern philosophy managed to prevent these episodes by furnishing the concept of reality with objectiveness. For example, the German notions of *Realität* (reality) and *Wirklichkeit* (effective reality) refer to two different realms. The first relates to the sphere of a concept onto which the subject is placed—i.e., the sphere of quality. The second belongs to the sphere of modality, that is, how an object exists as such—how it really exists. This is Kant 101. And yet, this transcendental bonding between subject and object through a normative plane of practical existence cannot guarantee that reality is necessarily real.

The conceptual status of reality was one of the main issues to deal with in the mature reflections of Hans Blumenberg—let alone his early drafts on the subject, like his first published article, entitled 'Atommoral,' or his 1956 draft 'Automation.' When it came to tradition, Blumenberg (2020a) went against the grain and turned the ancient dictum against the poets—that they were liars—upside down. What if they had told the truth? The German philosopher tried to take the consequences of such turn to the hilt. He meticulously reconstructed the history of the concept of reality—or at least its metahistory—and how its status was nothing but true. According to Blumenberg, 'this logical scheme does not necessarily coincide with the historical possibilities' (2020a, p. 501). Such possibilities would mean disenfranchized historical phenomena. Otherwise put: art—but also works in general—would have the chance to exist holding out the normative lattice of reality.

I would like to test these theses of Blumenberg by focusing on the historical status of current technology. I argue that technology—that is, the ongoing process of technicization (*technisierung*) over both life and existence—can effectively outstrip the normative, that is, the historical structure of reality. In fact, and as the reader can grasp almost elsewhere in Blumenberg's oeuvre, especially since the early 1980s, the question of technology and the question of reality became a single entangled singularity (2020b). Therefore, what theoretical tools are still available for a description of such state of affairs? Is there a phenomenological description—or even a metaphor—than can depict this unique non-historical phenomenon? These are the questions that this article aims to explore.

If technology (I am using this word as a broad translation of both *technik* and *technisierung*) was, according to Rüdiger Zill, one of Blumenberg's 'lifelong preoccupations' (2017, p. 291), then 'technology' becomes not only a 'paradigmatic' (Blumenberg, 1999, pp. 465–466; 1987, p. 103) concept but a gateway for essential philosophical problems. However, for the 'historian of technics' technology demands, first and foremost, a global survey of its reception and development. In his *History of the Spirit of Technics*, Blumenberg—after stating the exit (*heraustreten*) of technics from the dimension of history—reminds how '[t]he history of technicization suffers from a lack of language, from a categorial deficiency' (2009, p. 27). This conceptual insufficiency can be explained by the untuned link between modern self-consciousness—'pre-scientific' and 'pre-technological' (2009, p. 28), adds Blumenberg—and the dynamic, history-free pace of technology.

Such an unbalanced situation (Blumenberg, 2020b) responds to a previous, structural disposition of European consciousness. Up to now, the Husserlian scholar—and Blumenberg was, although with pronounced discrepancies, one, too—has been noticed on the striking similarities between this diagnosis and Husserl's three-leveled genesis of the sense of the 'original ark, Earth.' At the end of his 1934 manuscript, entitled *Foundational Investigations of Phenomenological Origin of the Spatiality of Nature*, Husserl (1981) distinguishes between a 'constitutive genesis' and its 'earthly precedence'—that is, the pre-scientific comprehension of the Earth as 'basis' and not as a Copernican 'body' among other 'bodies'—and the 'all natural-scientific knowledge of actuality and real possibility.' But while Husserl dates back this pre-scientific consciousness of the 'Ego' into the 'historicity' of every 'primitive' and 'cosmopolitan' people, Blumenberg –whom later called out Husserl for being 'proud' (*stolz*) about his 'minimum knowledge of the history of thought' (1997b, p. 85)—finds it in the 'antithesis between nature and technics' (2009, p. 26, 2020a, p. 363). Such a prognosis was even partially reconstructed by Blumenberg in one of his most challenging think-pieces, entitled 'An Anthropological Approach to the Contemporary Significance of Rhetoric,' where he argues that '[s]elf-persuasion underlies all rhetoric in external relations; it makes use not only of the very general, practically effective propositions [...] but also of self-understanding through self-externality' (2020a, pp. 207–208). This quote axiologically develops Blumenberg's description of the Earth as a 'cosmic exception' (1987, p. 679).

Nonetheless, Blumenberg's goal it is not to provide a boost for Husserl's 'anti-toxin of the technicization' that so terrified the latter (2020a, p. 388), whom since the beginning of his phenomenological program stormed out both psychologism and physicalism. On the contrary, Blumenberg aimed to point out how the 'highest degree of compression of our current scientific and technological condition'—a 'subject for itself, to the history of technology'—also represents 'an exposure' ['*eine Gefährdung*'] to the 'inexhaustibility of its progress towards new constellations' (2009, p. 84). Such a peril would end in a never-ending index of scientific and technological progress that even the most organized division of philosophical and intellectual labor could barely comprehend.

In *The Legibility of the World*, Blumenberg asserts how the 'conceptual emerge through the simplification of its outline,' allowing, thus, 'its recognition [*Wiedererkennen*]' (1981b, p. 22). How, then, this endless index of progress would enable a basic streamline of description-driven concepts? Rather than metaphors, an alternative accompanies Modern Age's self-conception—which it is, according to Blumenberg, a 'replica' rather than a mere 'antagonism' to Middle Ages intellectual foundations—and its consequent 'self- empowerment.' Such a systematic exercise of self-assertion can also perform as a response to a higher level of disorder. That is to say, Modern Age and its self-empowerment could aim at a global leveling of its comprehension ratio. So, another 'level of exit' [*Ausgangsniveau*] can be achieved through the commence of a 'new, constructive conception' (Blumenberg, 2009, p. 133). Otherwise put, our analog mindset it is not condemned to an endless feeling of impending doom regarding its online milieu. Rather, its self-awareness is also the recognition of its uninvited yet communal guest: technology. 'The antinomy of technics,' reflects Blumenberg at the end of his posthumous *Theory of the Lifeworld*, 'lies between achievement and comprehension' (2010, p. 171). Like a delayed and reverberating echo of the past—the outburst of the rhetoric language machinery—technology sonically assaults the self-conception of Modern Age. Its unflinching achievements irretrievably comes after the possibility of its comprehension. An asymmetry like this amplifies that dense assertion on technology—or technics as a whole—and its emergence out of history.

3 Exits

It is important to bear in mind that a significant amount of Blumenberg's considerations regarding the liaison between technology and reality are not easy to comprehend (2015a). While its difficultness is rather different than that of Heidegger's much undecipherable expressionism, developed in his meditations on technology (1977), Blumenberg's takes are based—just like Martin Meyer noticed in 1981—on a 'phenomenological anthropology' (Zill, 2020). Although the reader could not frantically believe that if one understands Blumenberg's account on technology the contemporary problems of technics would immediately be solved, like that Heideggerian enthusiast whom, according to Blumenberg (1997b), was just replacing his apocalyptic faith into a quasi-holy comprehension of contemporary world, a brief yet proper elucidation on his theses on technology could champion a more historically suitable concept of technics.

To what extent Blumenberg's contributions to this subject stands out among those of his contemporaries? The names of Ernst Kapp (2018), Richard N. Coudenhove-Kalergi (1922), Oswald Spengler (1963), Ernst Jünger (2017), Martin Heidegger (2017), Max Bense (1949), Günther Anders (2016), and Lewis Mumford (1967)—most of them familiar to Blumenberg, particularly the positions of Heidegger and Anders—immediately resonate at the moment of pondering such a colossal topic. While for obvious reasons I cannot even offer an overview to this problem, I will focus on Blumenberg's main thesis concerning technology. As it was already stated, his interest in both technology and its history it is intertwined with his works on the concept of reality, inasmuch as the reader can elucidate this by noting Blumenberg's initial perked up outlines on metaphorology, and then in his late description of the human being anthropogenesis. Regarding this historical co-dependence between history and technology, Helmut Müller-Sievers reminds us how 'in the deepest recesses of medieval theology and technology, the technological history of the intellect [*Technikgeschichte des Geistes*] and the intellectual history of technics [*Geistesgeschichte der Technik*] intersect and change into one another' (2012, p. 165). Nonetheless, at some point this manifolded nexus abruptly ceased as technology outbroke from history. 'Realities,' subsequently, become powerless concerning both its self-guidance and self-awareness. In *Cave Exits*, his last published book, Blumenberg has characterized Modern Age as 'the epoch of unlimited consciousness,' a historical plane upon which a forceful traffic between 'consciousness' and 'self-conception' becomes unlimited (1996, p. 12). The aforementioned sentence 'the unchosen visite of loss' enclosures such a situation. Since Fontenelle's otherworldly speculations (Blumenberg, 1981a), Modern Age's self-conception has been constantly swarmed by a bundle of disorienting forces. These forces have been episodically discharged, first and negatively, through literature (Blumenberg, 2020a), and then by the incessant mirroring of cutting- edge technology in cinema—let alone the current and naïve use of apps and telematic platforms. Thus, Blumenberg's aphorism: 'history is made against the weariness of its situations' (1998, p. 115).

The exponential autonomy of technics—a phenomenon that is scattered across Blumenberg's thought since the 1950s and more poignantly in his 1980s reflections—moves from a dynamic of existential stress, due to an over-technicized world, that later leads to a pandering living through the wellness granted by technics and its very fabrications, both created to soften and enhance a large set of ailments and human necessities. As Oliver Müller asserts (2008), 'the pathos of technic achievement' shows itself as the moment of 'self-realization,' a process that enables by itself new necessities. In sum, the emergence of technics is misinterpreted merely as an 'instrumental of technical service' (Müller, 2008). That Husserlian 'antitoxin,' which is an expansion of Blumenberg's succinct reflections on the concept of infection as an 'absolute metaphor' (1998, pp. 93–94), is meant to be inoculated not in technics but in the attitude with which technics and its history has become, when it comes to its comprehension, unavailable. If '[t]he result of the consistent development of the exact sciences in the modern period was a true revolution in the technical control of nature,' following Husserl's appendix to his *Crisis* (1970, p. 271), such a revolution becomes intelligible to the phenomenologist as long as its recognition as a kindred spiritual phenomenon remains as a blind spot.

Therefore, the dressing-down critique of Husserl regarding the loss of man's 'true being' (1970, p. 13) needs to be properly calibrated. In *The Laughter of the Thracian Woman*, Blumenberg stresses how '[t]heory leads people to become unnecessarily vulnerable to the world, to lower their guards, as Thales has done' (2015b, p. 28). Was Husserl's 'antitoxin' just another theory doomed to its own vulnerability, expressed through a searing desire that aimed at the taming of an untrue machinery? If the anecdote of the Thracian woman and its reception has shed light on the attitude of philosophy towards reality (Blumenberg, 2015b)—turning the title *The Legitimacy of the Laughter of the Thracian Woman* more philosophically fitting—the attitude regarding technics and its history reveals itself as the systematic self-defense of the concept of reality. This is what Carl Schmitt argued against Blumenberg's 'self-preservation' (*Selbstbehauptung*) as 'self-legitimation' (*Selbstermächtigung*) (Mehring, 2014). Regarding the book on Thales' anecdote, Schmitt called it 'a true masterstroke' (Lepper and Schmitz, 2007).

4 Reality

The concept of reality crafted by Blumenberg since his early academic career (Wertz, 2000) is lodged in Husserl's phenomenology. His insistence on the use of the plural *realities* and not a narrowed comprehension of *the* reality eschews from '[p]hilosophy's preference for language's semantic relation to reality' (Bluemenberg, 2020a, p. 180). This preference is, according to Blumenberg, the visible strata of a multi-layered 'absolutism of truth' (2001, p. 350; 2018, p. 3). In one of his posthumous writings published in the late 2000s, the Münster professor developed a thesis that he already had presented in 1964, namely, that the cut-to-fit concept of reality on which Modern Age has cemented both its self-conception and self- empowerment, presents an 'open context' [*ein* [...] *offenen Kontextes*] feature (Blumenberg, 2001, p. 362). This open, or, literally, 'public' context, strengths the undoubtedness of reality, a presumption which dynamic performs between global guarantees and a set of demands. The turmoil of any irreal experience triggers a requirement for the equilibrium of initial certainties. The Kafka sentence (Janouch, 2012) later quoted by Blumenberg, that is, '[t]he actual reality is always unrealistic,' suggest the ingenuous demands of those realistic claims that wish to outwardly deduct supposed threads of an inward reality. Amidst an unrealistic landscape and the economy of demands for reality—most of them placed in the decentralized concepts of seriousness, that is, ethics, morality, and politics—an unscathed 'absolutism of reality' (Blumenberg, 2018, p. 15) still capitalizes Modern Age self-comprehension. Yet even this self-comprehension is a technical aftermath. 'All is technics,' reminds Blumenberg (1996, p. 394) in his last book.

In his *Lebenszeit und Weltzeit*, Blumenberg rapidly refers—apropos the insurmountable rivalry between the lifetime and the worldtime—to the 'over-speeds of the technoid world' and its 'actualizations of the past and the upcoming' (1986, p. 27). Such a dislodged use of the concept of technics represents an exception within Blumenberg's work, as his systematic account on this subject was always executed in both an impassioned and meticulous fashion. Insofar, his contributions to the question of technics and reality had had focused on a prolific yet unorthodox approach to these subjects. However, in the context of the book of 1986, the sanctioned neologism 'technoid' is to be understood as the morose notice of a much larger, stark symptom. Blumenberg alerts about 'a pathology concerning time reference' [*einer Pathologie des Zeitbezugs*], a phenomenon that, first and foremost, expresses through time acceleration—Blumenberg had already underlined this in his reflections on rhetoric (2020a, p. 195). Our conventional use of the concept of 'rationality' manifests through an implicit 'benefit' of 'the element of acceleration, of the concentration of processes' (Blumenberg, 1981a, p. 410). Science, thus, understood as the successful agency of that 'absolutism of reality,' prompts such an acceleration (Blumenberg, 1981a, p. 410). This thesis is somewhat akin to that of Ludwig Fleck (1981) regarding 'conceptions,' and how they 'aspire' to the 'status' of 'logical systems.' Put differently: the fact that scientific conceptions aspire to become 'logical systems' demonstrate the duress of a greater rationality conducting accelerated scientific debates. In this perspective, it is quite intriguing that the name of the most conspicuous scholar of Ludwig Fleck, is Thomas Schnelle.

To Modern Age, therefore, rationality is the threshold where things can exist sooner. Hence, Blumenberg's definition of history as 'the separation of expectation [*Erwartung*] and experience [*Erfahrung*]' (1986, p. 66). If this is so, how to come to terms with the 'monotheism of reality' thesis and to the separation that give rise to history? In the context of Blumenberg's late reflections on the question of reality as the open access of time and its unquestioned inner logic of acceleration through iteration, the 'absolutism of reality'—the alien echo of the rhetoric machinery and its monstrous reverberations—is now grasped as the shielding process of recognizing the description of an unfathomable topic: reality as such. The increasing coefficient of self-consciousness depends on recognizing the unthinkable lifeworld upon which a description of reality is experienced as reality. Blumenberg (1986) uses a significant metaphor to depict this situation, that is, the amusement park. Both experience and expectation work hand in hand in this mundane facility. People expect exactly what they have been promised to, as they experience their own demands, that is, their expectations chained to fulfill their desires by default. Blumenberg's refurnished 'anthropology' (Hudson, 1993) would suggest, then, to question the apparent encompassing bond between science and reason, as the former needs the asynchronous analysis of the latter. Otherwise, just like Wittgenstein's fly, those smothered clients of the amusement park would die without experiencing the aging of their very subjectivity, without meeting the equilibrium—Blumenberg adds—between 'resignation, fulfillment, abandonment, and demand' (1986, p. 306). Was Blumenberg thinking in *Der Send* festival, which both him and his students were mesmerized by in the very summer semester of 1984 at the Münster Castle? Regarding this festival, Blumenberg exclaimed in the eighth lesson: 'I know, ladies and gentlemen, it is incredibly difficult to work your way through the broadcast today, to resist the temptations of the rollmops and everything else offer there. I almost got stuck, too' (Goldstein, 2020, p. 515). The theme of the seminar was entitled *Realität und Realismus*. It seems that the subject became available just after Blumenberg's leaving *Der Send*. What else could anyone have expected? Blumenberg—just like his students—did not leave the festival, but he escaped from it. That the human being 'does not make history,' but 'he sets the tempo of history' (Blumenberg, 1986, p. 242) explains the act of redress developed by people when it comes to establish coherence between time, reality, and experience. Thus, the image of the opened scissors between the lifetime and the worldtime. The shadow casted by a 'technoid world' on everyday reality makes one think if the total destruction of the lifeworld—an image sometimes mocked by Blumenberg (1997b), and then accounted as something quite serious (Blumenberg, 1986)—it is something unfathomable, or maybe the undeterred experience from which any reality is committed with reckless abandon. It remains an open question if the delayed self-conception of Modern Age regarding technics also shares Blumenberg's thesis on how this epoch 'decreed the value of everything prior to it as nil' (1985, p. 99). In this case, the outbreak of technology would dovetail to a more reluctant self-awareness concerning its past (Blumenberg, 2010).

I would like to conclude this general reconstruction of Blumenberg's theses on technology and reality with some remarks regarding our contemporary experience of the worldtime. To think that technological disasters would become more familiar if they could be linguistically forwarded to a series of metaphors, it is somehow a default assumption of an accelerated historical self-consciousness. This state of self-awareness presupposes a naïve immediacy between the undeducible plane of technology and a historically damaged, and temporally delayed, description of reality. Is this so?

In an interview conducted in 2007 by Jan Ross and Bernd Ulrich of *Die Zeit*, German Theologian and Cardinal Walter Kasper, and Cicero's most renown German intellectual of 2018, Peter Sloterdijk, exchanged reflections and remarks on the so-called 'return of religion.' The dialogue, entitled *Religion Is Never Cool* (*Religion ist nie cool*), addressed the negative potential of monotheistic religions and the supposed religious nature of human being. Regarding this cultural hotbed, Sloterdijk elaborates his non-religious view on human anthropotechnics under the metaphor of the reactor. Namely, how highly heated energies can be conducted and destructively practiced in societies depending on how much closer they are to the glowing core of monotheistic beliefs. Through such a proximity—and here the German philosopher strengths the analogy with a second metaphor—human being can access to 'unheard motivational sources' within himself (Ross and Ulrich, 2007).

It is quite curious that Sloterdijk's counterexample to Walter Kasper's canonical use of the historical paradigms of secularization and Modern Age self-liberation draws upon the image of a nuclear reactor. Three years before Fukushima and 20 years after Chernobyl, Sloterdijk's metaphor lucidly depicts that 'technoid world' to which Blumenberg referred to. On the one hand, the reactor does fully integrate the zeal of out of hand religious energies. On the other, these energies—and both metaphors as such—conventionally repeat the typecasting of both technology and monotheistic religions as contemporary culprits of destruction and malaise. Nowadays, one might think that the not so recent epidemiological crisis would have reprise the vocabulary of infections, toxins, and viruses. In this perspective, the number of essays, think-pieces, chronicles, and books published since early 2020, rapidly revealed the knee-jerk conceptual subordinance of their positions—let alone their homogenous degree of self-awareness. While this array of intellectual production mainly submitted to a public self-appraisal of previous thesis and diagnoses, it also demonstrated the still vexed relationship between ongoing *technisierug* and the much more endurable use of out-of-date concepts. Also, the fact that Xavier de Maistre's *Journey Around My Room*, Montaigne's *Essays*, Boccaccio's *Decameron*, and a large number of classics and not so classic titles—including Blumenberg's *Infection as an Absolut Metaphor*—became mandatory readings between 2020–2021, somehow confirms Blumenberg's remarks on reality. In one of his articles of 1964, he asserts how

> [w]e should not be surprised or misled by this historical phenomenon; it is quite natural that the most deeply bidden implication of an era—namely, its concept of reality—should become explicit only when the awareness of that reality has already broken' (1985, p. 99; 2020a, p. 512).

That Eavan Boland's *Quarantine* resonates so closely to such a global, murky experience—'In the worst hour of the worst season/of the worst year of a whole people' (2009, p. 282)—can be seen, according to Blumenberg, as a good example of how this depiction of reality is bounded to an already broken experience, from which one still remained unfazed.

In *Lebenszeit und Weltzeit*, Blumenberg ponders: "The most important is only done in the lateness, in the melancholy of the loss, although in the richness of the expression grown over time" ["Höchstrangiges wird nur in der Verspätung, erst in der Wehmut des Verlustes, aber im Reichtum des über die Zeit zugewachsenen Ausdrucks gemacht"] (1986, p. 194). Among cryptocurrencies hijacking the datasphere, tyranny-driven all-things AI, geopolitical eminences grises, and several other baneful postcards of the Pyrocene, one could easily take this aphorism of Blumenberg as a stark warning of truth. Both his birthday commemoration and the congresses organized around it prove this so. Either approaching historical chaos inside a vessel without a driver or displaying last minute security measures when facing an imminent explosion of conceptual energies within the technological reactor of language, one could rightly think that the experience of expectations and demands, despite how real or unreal can be, must become an active exercise of observation and awareness. Even if nobody's driving.

References

Anders, G. (2016). *On Promethean Shame*. In C. J. Müller (Ed.), *Prometheanism. Technology, Digital Culture and Human Obsolescence* (pp. 33–40). Rowman & Littlefield.
Bense, M. (1949). *Technische Existenz. Deutsche*. Verlags-Anstalt.
Blumenberg, H. (1981a). *Wirklichkeiten, in denen wir leben*. Aufsätze und eine Rede. Reclam.
Blumenberg, H. (1981b). *Die Lesbarkeit der Welt*. Suhrkamp.
Blumenberg, H. (1985). *Work On Myth*. MIT Press.
Blumenberg, H. (1986). *Lebenszeit und Weltzeit*. Suhrkamp.
Blumenberg, H. (1987). *The Genesis of the Copernican World*. MIT Press.
Blumenberg, H. (1996). *Hölenausgänge*. Suhrkamp.
Blumenberg, H. (1997a). *Shipwreck with Spectator. Paradigm of a Metaphorology for Existence*. MIT Press.
Blumenberg, H. (1997b). *Ein mögliches Selbstverständnis. Aus dem Nachlaß*. Reclam.
Blumenberg, H. (1998). *Begriffe in Geschichten*. Suhrkamp.
Blumenberg, H. (1999). *The Legitimacy of the Modern Age*. MIT Press.
Blumenberg, H. (2001). *Ästhetische und metaphorologische Schriften*. Suhrkamp.
Blumenberg, H. (2009). *Geistesgeschichte der Technik, mit einem Radiovortrag auf CD*. Suhrkamp.
Blumenberg, H. (2010). *Theorie der Lebenswelt*. Suhrkamp.
Blumenberg, H. (2015a). *Schriften zur Technik*. Suhrkamp.
Blumenberg, H. (2015b). *The Laughter of the Thracian Woman. A Protohistory of Theory*. Bloomsbury.
Blumenberg, H. (2018). *Rigorism of ruth. "Moses the Egyptian" and Other Writings on Freud and Arendt*. Cornell University Press.
Blumenberg, H. (2020a). *History, Metaphors, Fables. A Hans Blumenberg Reader*. Cornell University Press.
Blumenberg, H. (2020b). *Realität und Realismus*. Suhrkamp.
Blumenberg, H. (2021). *Quellen, Ströme*. Eisberge.
Boland, E. (2009). *New Collected Poems*. Norton.
Burnyeat, M., & Frede, M. (2015). *The Seventh Platonic Letter*. Oxford University Press.
Coudenhove-Kalergi, R. N. (1922). *Apologie der Technik*. Der Neue Geist.
Fleck, L. (1981). *Genesis and Development of a Scientific Fact*. The University of Chicago Press.
Goldstein, J. (2020). *Hans Blumenberg. Ein philosophisches Portrait*. Matthes & Seitz.
Haverkamp, A. (2012). *The scandal of metaphorology*. *TELOS, 158*(40), 37–58. https://doi.org/10.3817/0312158037
Heidegger, M. (1977). *The Question Concerning Technology and Other Essays*. Garland Publishing.
Heidegger, M. (2017). *Ponderings VII–XI. Black Notebooks 1938–1939*. Indiana University Press.
Hudson, W. (1993). *After Blumenberg: Historicism and philosophical anthropology*. *History of the Human Sciences, 6*(4), 109–116. https://doi.org/10.1177/095269519300600407

Husserl, E. (1970). *The Crisis of European Sciences and Transcendental Phenomenology. An Introduction to Phenomenological Philosophy*. Northwestern University Press.
Husserl, E. (1981). *Foundational Investigations of Phenomenological Origin of the patiality of Nature*. University of Notre Dame Press.
Janouch, G. (2012). *Conversations with Kafka*. W. W. Norton.
Jünger, E. (2017). *The Worker. Dominion and Form*. Northwestern University Press.
Kapp, E. (2018). *Elements of a Philosophy of Technology. On the Evolutionary History of Culture*. University of Minnesota Press.
Lepper, M., & Schmitz, A. (Eds.). (2007). *Hans Blumenberg/Carl Schmitt: Briefwechsel 1971–1978 und weitere Materialen*. Suhrkamp.
Mehring, R. (2014). *Carl Schmitt. A Biography*. Polity.
Müller, O. (2008). *Natur und Technik als falsche Antithese. Die Technikphilosophie Hans Blumenbergs und die Struktur der Technisierung*. *Philosophisches Jahburch, 115*, 99–124.
Müller-Sievers, H. (2012). *Kyklophorology: Hans Blumenberg and the intellectual history of technics*. *Telos, 158*, 155–170. https://doi.org/10.3817/0312158155
Mumford, L. (1967). *Technics and Human Development*. Harcourt Brace Jovanovich.
Ross, J., & Ulrich, B. (2007, February 8). *Religion ist nie cool*. *Die Zeit*. https://www.zeit.de/2007/07/Glaube-Interview
Spengler, O. (1963). *Man and Technics. A Contribution to a Philosophy of Life*. Alfred A. Knopf.
Wertz, A. (2000). *On the Possibility of Creative Being: Introducing Hans Blumenberg*. *Qui Parle, 12*(1), 3–15.
Zill, R. (2017). *Von der Atommoral zum Zeitgewinn: Transformationen eines Lebensthemas. Hans Blumenbergs Project einer Geistesgeschichte der Technik*. In A. Friedrich, P. Gehring, C. Hubig, A. Kaminski, & A. Nordmann (Eds.), *Technische Nichtwissen. Jahrbuch Technikphilosophie* (pp. 291–313). Nomos.
Zill, R. (2020). *Der absolute Leser. Hans Blumenberg. Eine intellektuelle Biographie*. Suhrkamp.

Gonzalo Forsberg, Researcher of Philosophy at the Instituto de Filosofía of the Universidad Diego Portales, Chile. The author of kamalatta (2017), Carl Schmitt's *Hamlet oder Hekuba* and the Question of a Philosophy of History (Leiden, 2024), and Rinomura (forthcoming), he has devoted his work in the fields of modern philosophy, aesthetics, phenomenology, poetry, and children's literature.

Unburdening the World with 'Worlds on Demand.' A Time Capsule from Hans Blumenberg's Legacy

Martina Philippi

Abstract Hans Blumenberg's estate contains an undated fragment entitled "On Machines and Simulators." In this vision of the future, Blumenberg states that there is one type of machine that predominantly produces material things and another that is similar to consciousness and saves us time. This has the fatal consequence that by means of the second type, we have a lot of time to consume the material products from the first type, so that these products 'clog up' the world. Blumenberg's proposal is to have machines produce immaterial products for consumption instead, namely 'simulators' that create 'worlds on demand.' In this entertaining short text, some aspects of Blumenberg's work emerge that are ideally suited to a theoretical understanding of the phenomenon of digitalization and its consequences. The article compares Blumenberg's vision with the current state of affairs and, on the basis of current phenomena, draws on the theoretical strands from Blumenberg's work that are alluded to in the fragment. To this end, it asks what has become of the idea of 'worlds on demand' and explains the process of the digital transformation of the life-world as a further development of the emergence of virtual worlds. From here, considerations from "Phenomenological Aspects on Life-World and Technization" and *Lebenszeit und Weltzeit* take on a new meaning. Finally, an outlook is given on the extent to which this updating of Blumenberg's phenomenological anthropology can be related to current challenges of technology.

1 Introduction: A Blumenbergian Futurologist Fragment

To reflect on history does not exclusively mean taking a look backwards. It is also an opportunity to have a look in the future, to create a future scenario. Those future scenarios can be read at a later date as a special kind of historical testimony, because every age of time has its own anticipated futures, its own futurology. And they do

M. Philippi (✉)
Faculty of Arts and Humanities, Universität Paderborn, Paderborn, Germany

A. Fragio et al. (eds.), *Hans Blumenberg's History and Philosophy of Science*, Studies in History and Philosophy of Science 63,
https://doi.org/10.1007/978-3-031-78251-0_9

not only relate to anticipated technologies in an isolated sense: instead, a whole world is imagined in which particular technologies are embedded in everyday routines and practices. Since introducing new technologies always affects life in a holistic way, such scenarios are very revealing when we integrate novel technologies into our life-world.

For Blumenberg, the history of technology is also the history of science, in the sense of a historical clarification of the relationship between theory and the life-world. A central text that can be read in these terms is "Phenomenological Aspects on Life-World and Technization," first published in 1963. As Blumenberg states in this text, it is the embedding in the life-world that gives technology the strange peculiarity of becoming invisible in its use: "The technical as such becomes incomparably more invisible by being implanted into the life-world than through the mimicry of casings and covers" (2020, p. 386).

This consideration can also be applied to technologies in future scenarios: to assess novel technologies means to look at them in their possible contexts of application, and therefore at least as interesting as the question of *which* technologies will be available to us in the future is the question of *how* this will change the way we live. Therefore, the phenomenon of future scenarios coming true (or not) provides insight into historical development in a comprehensive sense. In the retrospective, we see not only what has become of those anticipations, but also under which circumstances and conditions the development has taken place. From this perspective, the very needs become visible that led to the realization of technical visions, and thinking further, we also see the changes, chances and challenges that emerge from the respective new technology.

An extreme example of future scenarios is dystopian science fiction where concerns about the future are transformed into a story. But science fiction in general has long proven to be a creative and fruitful approach to reflecting on the history of science and technology. Reading classical science fiction novels or watching science fiction films produced in former times, means observing the technical realization of visions from the past when the once anticipated future has arrived. Some scenarios designed for a distant future can now be technically realized,[1] and how it actually turned out can be compared with how people in the past thought it *could* turn out. So, in retrospect, we see what has become of the anticipation, and in addition, we see under which circumstances and which conditions that has happened. We also see what needs have arisen that led to the realization of technical visions, and what might emerge from that development. Looking back on visions from the past gives us a contrasting foil against which we can observe current developments and learn to understand how they came to be real.

[1] For example, the idea of the holodeck from Star Trek can be found in virtual, immersive environments that make it possible to experience simulated environments through VR glasses or as a virtual environment. Visions for the automated sensory recognition and identification of people—such as in *Minority Report* via panels or small, fully autonomous search robots—or new possibilities for voice control—such as the talking directly to the computers in Star Trek—have left the sphere of science fiction and have become part of applied systems as developed technologies.

So, observing science fiction coming true implies some valuable perspectives. This also applies to future scenarios that are part of theoretical considerations. An undated fragment by Hans Blumenberg, entitled "Über Maschinen und Simulatoren" ("On Machines and Simulations"), has been posthumously published in the volume *Schriften zur Technik*. In this small text, Blumenberg ventures a look into the future and anticipates technical developments—with a certain interest. In the first sentence of the fragment, he states his idea: "We form borderline ideas of the development in which we find ourselves in order to clarify its tendency" (Blumenberg, 2015, p. 298). The notion of borderline concepts is nothing entirely new in his thinking: similar to the 'borderline concept' (such as 'reality' or 'life-world'), such borderline ideas indicate something that is not (yet) definable but can be guessed at; they occupy a functional void in considerations to enable tentatively thinking about them (cf. Blumenberg, 1979).[2]

In my contribution, I would first like to present the 'borderline idea' from Blumenberg's fragment where he develops a scenario of 'worlds on demand' that are supposed to 'unburden the world' from an overwhelming material production that is driven by our willingness to consume. Then I would like to ask what has become of the 'worlds on demand' now, i.e. what this 'frontier idea' has developed into. Finally, I will suggest reading Blumenberg's fragment as a time capsule he has left us in his legacy. A time capsule is typically an object that preserves its content for later discovery by future generations. In a metaphorical sense, it can also be a piece of theory that unfolds a special impact in the future, namely against the background of knowledge of a later time. As in other small texts by Blumenberg, certain concepts and thoughts allude to larger complexes of his thinking (cf. Philippi, 2010): terms such as 'unburden' or speaking of 'realities' in the plural form are reminiscent of his reflections on the absolutism of reality or phrases such as 'realities in which we live' and form a direct link between the light-footed text miniatures and the elaborate major works. In this sense, this fragment's light-footed, experimental form also contains direct paths into Blumenberg's thinking alongside a vision of the future. By reading the fragment as a time capsule, I suggest that it enables us to compare Blumenberg's future vision with our present to gain a different perspective; and at the same time reach direct access to Blumenberg's thinking as it could be applied to our current situation. So let's open the time capsule!

---

[2] On the notion of the 'borderline concept' and its usage, see Feger (2020), Klein (2017), and Niehues-Pröbsting (2022). It is a Kantian term that can be declared as follows: "Kant understands a borderline concept to be a concept that […] leads to the limit of our comprehension, in that it posits something conceptually without being able to determine it qualitatively, positively" (Eisler, 1904, p. 413, my transl.).

2 'Worlds on Demand': A Fragment and Its Vision

In the one-page fragment entitled "Über Maschinen und Simulatoren," Blumenberg announces to "form borderline ideas of the development in which we find ourselves in order to clarify its tendency" (2015, p. 298).[3] This approach is guided by a specific question: "What will become of the machines?" (2015, p. 298). Blumenberg then develops the following scenario:

> Let us assume that the world consists of two classes of realities, of objects and subjects. Then there would ultimately also be two types of machines: 1) those for the production of objects, 2) machines analogous to consciousness (i.e. machines that behave and perform as if they had consciousness, as is attributed to subjects) (2015, p. 298).

Their coexistence proves to be fatal because, according to Blumenberg, the machines of the second type save us time, which we invest in consuming the former's products. That leads to a problem that Blumenberg calls 'overpopulation' of the world:

> These two types work [at each other's hands] in a murderous way: the machines analogue to consciousness relieve people of almost all their tasks and free up their time and energy for any consumption of objects; the production machines supply these objects in any quantity and design.
>
> But: this world is already heavily populated, probably overpopulated, and in it, there is now to be room for all those objects constantly produced from automata, as they already clog up all the streets and squares [...]. The disadvantage of producing machines is that they produce real objects. In this, they are completely antiquated (Blumenberg, 2015, p. 298).[4]

Blumenberg formulates a solution that consists of the development of 'world simulators':

> We must subdivide group 1 and also allow machines that produce unreal objects. We call them simulators. We already have space simulators with which the conditions of weightlessness can be produced – but what we need and will have is not the space simulator, but the world simulator, which does not produce objects, but simulates them as a complete 'world'. It would be an advanced television set, for example. Colour and spatial television are already there, but never in such a way that the viewer can do more than watch (2015, p. 298).

Anyone who played a lot of Gran Turismo or FIFA (both game series that started in the 1990s) during the pandemic lockdown 2020 will recognize the following scenario:

---

[3] My transcription and translation.

[4] The original passage: "Diese beiden Typen arbeiten in mörderischer Weise einander [in die Hand]: Die bewußtseinsanalogen Masch[inen] entlasten die Menschen von fast allen ihren Aufgaben und geben ihnen Zeit und Kraft für einen beliebigen Konsum von Objekten frei; die Produktionsautomaten liefern diese Objekte in beliebiger [...] Menge und Ausführung. Aber: diese Welt ist ohnehin schon stark bevölkert, wahrscheinlich überbevölkert, und in ihr sollen nun auch noch all jene ständig aus Automaten produzierten Objekte Platz finden, wie sie schon heute alle Straßen und Plätze verstopfen [...]. Der Nachteil der produzierenden Maschinen ist, daß sie reale Objekte produzieren. Darin sind sie völlig rückständig."

> The world simulator will allow each participant to order a car of any model, get in, operate all functions, steer it in any direction, etc. – all without leaving their own four walls, but with the completely adequate pleasure that only reality has offered until then. You can order a stadium with a football match and programme your team's victory or leave it uncertain – depending on your personal preference (Blumenberg, 2015, p. 298).[5]

The conclusion then follows in the tradition of Blumenberg's anthropology: "The relief of man through technology is followed by the world's relief through perfect [sic!]. Everyone gets their favourite world delivered right to their door on demand" (2015, p. 298).

This fragment provides us with a forecast for further technical developments, which we can now compare with our time and our knowledge of the current state of affairs. This forecast is linked to some interesting motifs that appear quite familiar today: there is the idea of artificial worlds we can experience as if they were real; there is the fear that our technology-supported production of material objects could 'clog up' the world; and also the distinction between consciousness-like machines and those that produce material objects.[6] At first glance, this may sound only vaguely like what Blumenberg deals with in his elaborated works on anthropology, phenomenology, and philosophy of technology. But this impression changes when one takes some other texts from his work into account and links them to a closer look on the phenomena sketched above. Elements of Blumenberg's theories then clearly resonate, and we can connect the examination of current phenomena to them. So, in the following section, I will ask what has become of the developments whose tendency Blumenberg has tentatively outlined here.

3 What Has Become of the Machines?

What has become of the machines, the 'worlds on demand', and the relation of material production and consumerism? Trying to explore these questions by drawing on observations of current phenomena and processes of change can only be a sketch here, similar in form to Blumenberg's fragment. But by doing so, it is possible to relate Blumenberg's borderline idea of the development in which we find ourselves in his time with how things have developed actually—not with the

[5] In the epilogue of the posthumous volume *Schriften zur Technik*, another unpublished fragment is quoted in this context: "We will have to and can follow Kafka's advice to stay at home" (UNF 2955, quoted in Blumenberg, 2015, p. 297).

[6] Blumenberg's typification of machines is slightly reminiscent of Gotthard Günther's machine theory from 1976: "In the context of a philosophy of technology, we can therefore say in general terms that man has so far created two fundamentally different ideas of the machine in his technical development. The first is the classical Archimedean machine, whose purpose is to produce work. This has been supplemented by the idea of the 'second' machine, which is no longer expected to produce work, but information. The 'first' machine was designed in analogy to the human arm (and hand), the second is expected to be a technical reproduction of the human brain" (Günther 1976, pp. 96–97). I would like to thank Nils Kasper for drawing my attention to this parallel.

intention to showing whether he was right with his prognosis, but to connect his thinking to the current situation.

Today, we can say that we have got the 'worlds on demand' that Blumenberg requested in his fragment. The desired worlds are virtual environments that can be experienced with different senses, in which you can drive, fly, surf or wander through fantasy sceneries. Indeed, there are games in which you can 'order a car of any model' or program the settings for a soccer match. There are also simulated training environments that, like the space simulator mentioned by Blumenberg, make it possible to experience movement in a virtual space under certain conditions.

How we experience the impressions of such a world has now been given the title immersion: a deceptively real-looking simulation of real perception that is adapted to our perceptual habits and is technically implemented using various means, including VR glasses, projections, various types of controllers that resemble objects from specific contexts and devices that we can place ourselves in and that then place our body in a spatial position and subject it to certain movements that match the visually and aurally simulated world.[7] Furthermore, the worlds on demand are not only 'complete,' as Blumenberg insists, but also closed, and self-contained. They differ from the real world in the sense that we can enter them and leave them again,[8] in other terms: they provide experiences at the touch of a button that can be started and switched off again as desired. Gaming worlds and computer simulations, as we know them today, are virtual in a narrow sense.

That has some phenomenological implications. According to a phenomenological definition, a world is only a world if someone is there to perceive and experience it. Consequently, the immersive technology is explicitly designed for the perceiving subject. As in Blumenberg's vision of an 'advanced television set,' a kind of theatre is played out to this subject, which it can shape via settings or situation-immanent actions. While the worlds on demand must be variable in their details so that we can choose different car models or determine the outcome of a game, we are the only constant in them. What makes a world what it is, namely that it can be experienced, is precisely the disposition of these worlds on demand, which is only realized when we immerse ourselves in them, when we bring our expectations and our knowledge of how to do things in the real world into them. These worlds are interactive replicas of our environments and practices. We can move around in them as we already know how; as they are adapted to our real routines of perception and action, the effort required to adapt to them is felt to be no greater than in real situations.

---

[7] In this way, the simulation also includes kinaestheses in the sense of Husserl. Husserl's demarcation of the sensory life-world from the mathematical world of ideas gets an interesting twist here, because the two come together in the digitalized models of the life-world. After all, the appeal and power of these artificial worlds live from their—simulated—sensory qualities.

[8] There have been attempts to enrich some of these game worlds with functions that also have meaning in the 'real world,' e.g. Second Life with its possibilities to connect to real social functions. Online games such as World of Warcraft open up a further dimension of shared virtuality. Nevertheless, they are still games.

I would like to go deeper into these considerations in order to show that the problem of the 'clogging up' of the world by material objects has not been solved with the delivery of 'worlds on demand.' Blumenberg's idea of 'worlds on demand' is intended to decrease material consumption leading to a 'clogging of the world with objects.' We should consume worlds instead of products—and preferably stay at home. In this way, the world will be relieved of us, and that also means our consumption's consequences. But one could even say that this development has exacerbated the problem, i.e. that the establishment of virtual worlds and their widespread availability has led to an even growing 'overpopulation of the world' in another sense, namely in terms of meaningfulness.

4 Artificial Worlds and Consumerism

In his monograph 'The Day the World Stops Shopping', J. B. MacKinnon (2021) makes a very similar point: our world is collapsing under our consumption; if we want to maintain our consumer behaviour, we need to move it into cyberspace. Ideally, we will someday give it up completely. But until then, at least we do less harm by not consuming materially—very similar to Blumenberg's proposal, except that one can actually make financial transactions in MacKinnon's simulated worlds. So it's a kind of transitional stage designed to solve the problem of overconsumption.[9]

The question is, of course, why we consume so much that it threatens our livelihoods. MacKinnon cites economic (and political) encouragement as one of the reasons for our increased consumer behaviour. However, there are incentives to consume even without this encouragement. The thesis I would like to put forward here is that the incentive is the experience of an artificial world and its benefits, which, remarkably, can take place in the material world.

It is not the case that material consumption is simply the consumption of goods. Rather, it is also the consumption of possibilities, experiences and meaningfulness. Our consumption of material objects has also taken on traits of virtuality. Contemporary and older commercials can be seen as a marker of this development. Advertising has always worked by making promises beyond the product and its immediate benefits. But the nature of these promises changes over time. In recent decades, a product usually promised its buyer status, success, and attractiveness.

Old commercial adverts show us how buying a product has been supposed to change a person's image. You could become more attractive by using a certain shower gel, more successful by wearing a certain accessory, and more independent by driving a certain kind of car. That may sound trivial, but for some years, we have seen how the promises change while the products remain more or less the same. It is no longer so much about how a product changes a person, but how it affects how

[9] Cf. Chapter 19 in MacKinnon (2021): "Now we're all shopping in cyberspace?", pp. 249–258.

the person perceives and experiences things. In current advertising practice, the purchase and use of products provide a wider range of experiences: by consuming products, we open up new worlds and spaces of perception. A certain form of perception is now linked to the advertised products, which until then has been reserved for an explorative approach to the world. With a shower gel, we not only cleanse ourselves, but now 'experience' the freshness of a spring morning, and with a TV streaming subscription, we do not simply entertain ourselves, but 'discover' something new.

This is not necessarily surprising, as the concept of 'worlds' is very convenient to marketing. You can sell products better if you embed them in the existing life-world, i.e. if you suggest what you can do with them and what kind of impact they will have in your life. Doing so, personal perception can be influenced towards certain aspects of the offered good, for example experiencing in a certain way and carrying out certain predetermined courses of action. The worlds of experience have been designed by human actors based on our interests to guarantee their success; they are prefabricated to 'serve' our interests so that we select and purchase products we probably like.[10] Even the act of consumption has been given a strong experiential structure: the acquisition of material products now takes place in specially created shopping worlds, even worlds of experience.

How buying products can turn into consumerism, developing a special attitude not only towards the products, but also towards oneself and the meaningfulness of one's life, is described in John Carroll's essay "Shopping World".[11] In this essay, he takes the name of an American chain of shopping centers literally to show how the experience changes in such an artificial 'world of experience'. With this analysis in mind and contemporary consumer behavior in sight, one can say: The human capacity to create artificial worlds is instrumentalized by economic agents, carried out by marketing with some psychological basic knowledge. These artificial worlds fuel consumerism and try to shape our way of experience, our expectations toward the world and our priorizations—what is important right now, what is important in life—until today: they try to make us believe that everything we could want is available for money, no matter of the consequences. In other words: From the worlds on demand emerges a novel attitude of a demanding expectation towards the real world. And these consequences, which can be seen as a 'burden' for the world and humankind, are initially of a material nature, i.e. they concern resources and the 'clogging' of the world with things mentioned by Blumenberg.

These consequences are related to the technical conditions of digital artificial worlds. Technically, the creation of virtual worlds is based on two conditions. One condition is providing a material basis for the operation of simulation production machines. A wide range of complex technologies are required to make these

[10] Under the title 'personalization', this has become a marketing strategy that is increasingly being pursued today.

[11] Cf. Carroll 1980, pp. 95–105, especially p. 101: "Buying's function at Shopping World is to provide a context for exciting experience, in fact for experience that is quite opposite to the useful, dutiful collecting of provisions."

experiences possible at the touch of a button. Simulation software, graphics, sensor technology and even artificial intelligence are the building blocks of powerful simulation systems that require high computing power to process data quickly. Computer simulations are given even more possibilities through artificial intelligence. In this way, consciousness-like machines are now also involved in the production of unreal objects, which inevitably entails the production of further material objects.

The other condition is the processing of our own physical and mental experience by the world simulators, which goes hand in hand with the production of complex simulations. Because to interact with us, technologies need information.[12]

This requires processing everyday knowledge about us and our world into data that machines can process. Quantifying and categorizing models play just as central a role here as the automation of computing processes. Automation, in turn, is itself based on several technical and theoretical conditions. These include the fragmentation of workflows into processes, which in turn are optimized and standardized; the setting up of trigger mechanisms or triggers; and a decision-making authority that decides when which process is started and how, either in the sense of a procedural chain of command or, more sensibly in the case of complex processes, as self-learning artificial intelligence. And finally, a powerful infrastructure must be available, because data transfer and data processing in real-time is only possible based on networking.

This has two main consequences: Firstly, the 'consciousness-like' machines, i.e. AI, are also involved in artificial worlds, for example in analysing the data or designing the worlds according to personal preferences or other personal characteristics. Blumenberg's division of machines from the manuscript must therefore be modified from today's perspective. On the other hand, the technical conditions of artificial worlds impose high demands on our environment: for instance, an energy-intensive infrastructure and technical components that have to be manufactured, replaced and disposed of, i.e. burdens that are added to the immediate technological consequences and risks that affect data protection, security and user autonomy. The creation and experience of artificial worlds, as we receive them on demand today, therefore result in a further material burden on the world, and not relief.

In the following, I would like to show that these material burdens are not the only ones and that a 'clogging up' or overload of our life-world is intensified in another way: the overloading of consciousness with possibilities.

[12] Data collection is increasingly done by sensors and not by manual data transmission. Mark Weiser (1991) analyzed this linking of data, sensor technology, computing power and the real world in his famous reflections on 'ubiquitous computing.' Here, too, a similar effect occurs as when reading Blumenberg's text or other versions of the future from the past: handy, networked devices and sensors that make the manual input of data superfluous seem very familiar to us today.

5 The Digital Hybridization of the Life-World

The assumption that the availability of 'worlds on demand' has led to a change in the consumer experience and the expectations placed on it is import to answer an elementary question: Under which cultural circumstances are we willing to engage in artificial worlds like the 'worlds on demand'? To refer once again to the sentence from the "Phenomenological Aspects on Life-World and Technization": "The technical as such becomes incomparably more invisible by being implanted into the life-world than through the mimicry of casings and covers" (Blumenberg, 2020, p. 386). Then, how was it possible for us to become so familiar with immersing ourselves in these automation-based artificial worlds? What was the condition for Blumenberg's observation about technologies becoming invisible—and that means getting the characteristic of obviousness *(Selbstverständlichkeit)*—to the extent that entire artificial worlds emerge from them, and we are prepared to immerse ourselves in them and not only enjoy their advantages but also accept their side effects?

This central question affects reflections from the history of science as well as the history of society and technical innovation. Phenomenologically speaking, integrating technical structures of meaning into the life-world has the characteristics of an influx. Also beyond phenomenology, the phenomenon of the influx of scientific and technical structures of meaning into the life-world is nothing new. In his monograph *Patterns*, sociologist Armin Nassehi (2024) describes how the processes of digitalization are not an exclusive phenomenon of the computer age but build on pre-digital developments and trends and absorb them. The differences between the still very analogue world of Husserl and Blumenberg and our world, amongst the so-called digital transformation, are gradual. The willingness to integrate technology into the unquestioned fabric of the everyday world was also there earlier. There were issues of acceptance in the past as there are today, as well as the overly thoughtless use of technologies with serious consequences and the critical voices associated with them. But today we experience a development that affects our lives stronger than ever before, with more benefits and amenities, but also more undesirable side effects that are sometimes even hard to detect.

A hybridization between virtual and real worlds is taking place, enabling and encouraging us to engage in the virtual worlds with real consequences.[13] That also affects the constitutional characteristics of worldliness. Since world is always something that we shape in constant interaction, we cannot separate the 'artificial worlds' from 'the real world'; and since the virtual world is actively shared through networking and interweaving with our everyday world, collective participation is no longer a unique selling point for 'reality'—and this way of speaking has also lost its

[13] Even when we move through such worlds in a playful way, they can have a binding character. This is all the more true in artificial worlds that are not just game worlds, such as immersive web technologies.

To go deeper into this cf. Quent and Plattner (2024) on the responsible shaping of immersive digital worlds.

ground since, for example, cryptocurrencies have become economic and internet communication can have social and psychological effects of the most severe degree. The boundaries between the 'artificial' and 'real' world are blurring; our life-world is permeated by digital structures of meaning.

By 'structures of meaning,' I mean the options of interaction presented to us by digital technologies. These can be 'apps,' i.e. software, as well as 'smart devices,' i.e. technical objects equipped with sensors and a highly automated functionality. We are dealing here with material and immaterial objects that have in common that they act in a 'consciousness-like' way: we do not encounter them in our world as mere objects, but they actively address us—even if there is no AI behind them. These structures of meaning operate in the sphere typical of our intellectual life: the realm of pure meaningfulness. Being involved means engaging with them, allowing them to become meaningful and binding for us. Remarkably, we engage with these offers even though they are 'merely' virtual; they are not closed world simulators, but simulations of sections of the world; their role lies above all in the fact that they open up worlds. The somewhat out-of-fashion title 'portal' emphasizes this role particularly vividly: interactive websites, for example, are portals into a representation of a real section of the world.

Our relationship with those hybrid worlds is complicated. We bear responsibility not only for the technologies we develop, but also for the structures of meaning that we create with them. But this responsibility exceeds our capabilities because even these artificially created worlds confront us in a way that Blumenberg has called the 'absolutism of reality.' The development can be roughly described as follows: in order to overcome the absolutism of reality, we put myths into the world; our myths learnt to walk, they learnt to lead an independent life, and we began to enter into them, not only to live with them but also to live within them. The structures we set in motion developed their own dynamics; the automatisms we built began to confront us no longer as tools, but as partners in interaction.

These structures of meaning that we have created are different from Blumenberg's 'worlds on demand': they are not closed; they are unpredictable; they demand our attention; not only the concrete technologies, the mobile phone, the computer, the television, the smartwatch demand our attention, but also the practical requirements and possibilities they create: the technical necessities—charging the battery, software updates—, the so-called 'user competence,' the calls for action based on the respective technology. In many cases, we interact with these technologies—websites, apps, or devices—in a binding way, with real effects in a social or legal-contractual sense. They demand information, user expertise and procurement efforts from us and offer us social relationships, goods, experiences or options for action in return.[14] The worlds we have created have the potential to overpopulate the 'real world' with the multitude of different special settings, which, as we know, all have

[14] We can leave a large part of the maintenance work to automation, and we can familiarize ourselves with new software relatively quickly due to strong aesthetic standardization, but this only shifts the problem.

their own self-evident characteristics. Above all, however, they require us to utilize anthropologically relevant resources: involvement and attention.

6 Applied Blumenberg: Living in (Virtual) Realities and the Opening of the Time Gap

The digitalisation and hybridisation of the world we live in shows us what it means not only to live in a world *with* technology, but also to experience worlds that are technologised in the narrowest sense, that means created and shaped by technology. What we can take out of the time capsule of this small text, relates contemporary phenomena with two trains of thought from Blumenberg's work: first, the possibility and attractiveness of (technically created or supported) artificial worlds that relate the concepts of life-world and technization to each other, and second, the phenomenon that Blumenberg calls the 'opening of the time gap' in *Lebenszeit und Weltzeit*.

To do this, I will take up the above question once again: Where does our willingness to live in artificial worlds in the first place come from? As shown, through a certain culture of consumption, we have become accustomed to experiencing artificial worlds, which can be like fantasy worlds or daydreams, but are provided to us by others. We have become familiar with consuming artificial worlds of experience. This can simply be seen as a contingent phenomenon of economic and technological history. At the same time, however, there are anthropological aspects that go deeper than the economic interest in offering such worlds and the interest of their target groups in adopting them. We can approach these aspects with the above-mentioned theoretical approaches from Blumenberg's work. There we see a desire to create and design artificial worlds to cope with the contingency in which human life always stands; and we see the urge to close the time gap at least to some extent. I would like to show how the attempt to realise these two wishes can result in a constellation where the two, as described in Blumenberg's manuscript, 'work at each other's hands in a murderous way'.

The hybridization of the life-world by the influx of virtual worlds works so well because—as mentioned above—we do not only abandon ourselves to the products ('the technology') but also to entire worlds into which we 'immerse' ourselves, both with the same attitude of unquestioningness. The question raised above—where do the need for a technologized world and the willingness to live inside such a world and even adapt to it, come from?—can be approached by a piece of Blumenberg's theory on the relationship between theorization, technization, and the life-world. In "Phenomenological Aspects on Life-World and Technization," Blumenberg provides an explanation that follows on from Husserl's account of the modern history of science and simultaneously draws on basic scholastic concepts: "The emergence from the life-world, that is, from the 'universe of obviousnesses', was [...] the reversal of all of reality's characteristics of self-evidence into contingency.

Contingency means the appraisal of reality from the standpoint of necessity and possibility" ("Das Heraustreten aus der Lebenswelt, also aus dem 'Universum der Selbstverständlichkeiten', war […] Beginn […] des Umschlagens aller Selbstverständlichkeitscharaktere der Wirklichkeit in die Kontingenz. Kontingenz bedeutet die Beurteilung der Wirklichkeit vom Standpunkt der Notwendigkeit und der Möglichkeit her") (2020, p. 395; 2015, p. 198).

Blumenberg's considerations on the absolutism of reality as well as nihilism are echoed in the following presentation:

> The awareness of reality's contingency is, however, the foundation of a technical attitude toward what is given: if the given world is only an arbitrary segment from the infinite scope of the possible if the sphere of natural facts no longer exerts a higher justification and sanction, then the world's facticity becomes the nagging impulse not only to judge and criticize the real from the perspective of the possible but also to augment merely factual reality by way of realizing the possible, of exhausting the latitude of invention and construction, to form a cultural world that is in itself consistent and can be justified through necessity (2020, p. 395).[15]

You could say that if we have to live with the absolutism of reality, then at least we have shaped reality ourselves. And we fundamentally shape reality not only with the objects to which we consciously turn but above all with the conditions of daily life that we create ourselves—with all their practical requirements. This kind of self-created obviousness necessarily follows on from that which we already find in our everyday lives and our theoretical practices, and is extended by the creation of artificial worlds.

Blumenberg's understanding of technology implies that technology, which seems to fit smoothly into our everyday lives through automation and virtualization, not only expands the ground and horizon of our practice in the sense of a phenomenological epistemology by relieving our limited attention; but also fulfils the supporting function of the self-evident, and it expands our possibilities based on what we are capable of.

The process of technization raises the ambivalence of self-evidence, which characterizes our practices in the life-world, to a higher level. In "Phenomenological Aspects on Life-World and Technization", Blumenberg highlights this connection between the fundamental self-evidence of the life-world and that on the higher, technologized level as follows: "The process of concealing what has been discovered, as analyzed by Husserl, reaches its telos only when what has become unconscious through theoretical inquiry returns to being unquestionable" ["Der von Husserl analysierte Prozeß der Verdeckung des Entdeckten erreicht erst darin sein

[15] "Das Bewußtsein von der Kontingenz der Wirklichkeit ist nun aber die Fundierung einer technischen Einstellung gegenüber dem Vorgegebenen: Wenn die gegebene Welt nur ein zufälliger Ausschnitt aus dem unendlichen Spielraum des Möglichen ist, wenn die Sphäre der natürlichen Fakten keine höhere Rechtfertigung und Sanktion mehr ausstrahlt, dann wird die Faktizität der Welt zum bohrenden Antrieb, nicht nur das Wirkliche vom Möglichen her zu beurteilen und zu kritisieren, sondern auch durch Realisierung des Möglichen, durch Ausschöpfung des Spielraums der Erfindung und Konstruktion das nur Faktische aufzufüllen zu einer in sich konsistenten, aus Notwendigkeit zu rechtfertigenden Kulturwelt" (2015, pp. 198–199).

Telos, daß das im theoretischen Fragen unselbstverständlich Gewordene zurückkehrt in die Fraglosigkeit."] (2020, p. 386; 2015, p. 190).

By digitizing and hybridizing the life-world as the basis of their thinking and lives, humans might be once again turning away from their capability and also their need for reflection, which actually—in terms of theoretical curiosity—initiated and enabled the development of ambitious technologies in the first place; but this time, however, this turning away is deliberate, autonomous and definitive, anchored in life-world practice and the closed interaction with machines under predetermined, standardized options.

As Blumenberg states in his manuscript, there is a 'overpopulation' of the world by the products of the machines that turns out problematic. But even if this could by solved by virtual consumerism, another kind of 'overpopulation' manifests itself in the scarcity of a non-material resource: attention. This scarcity also manifests itself socially in the growing need for comprehensibility and orientation, for immediacy and focus—things that were perhaps not so important to us before, perhaps even taken for granted.

In this way, we see two effects of the production of virtual and especially hybrid worlds on the real world. On the one hand, the virtualization and hybridization of the life-world are exacerbating environmental problems that have become urgent and the scarcity of resources. On the other hand, there is another aspect: what was originally developed as an extension of human range is now proving to be a new anthropological imposition. I would like to explain this with Blumenberg's metaphor of the 'opening of the time gap,' which can be summarized as: too little time for too much world.

In the manuscript presented at the beginning, one can see how the time released by technology is consumed by technology again. It is not the case that the opening of the time gap, as Blumenberg describes it in *Lebenszeit und Weltzeit*, is cancelled out by technology; it is not the case that we use the gain in time made possible by 'machines' to open up a little more of the world than it was previously possible for us. Rather, we are filling the time frame that we have opened by using technology again with the usage of technology. So the time gap is widening even further because the possibilities are growing (while our lifespan remains the same).

In the pre-digital era, suffering from the time gap probably mainly affected those whose lives were largely driven by theoretical curiosity. As shown by Blumenberg, the contrast between 'theory and practice' becomes obvious in the anecdote of the laughing Thracian woman, which may have gained its popularity precisely because it illustrates the supposed irrelevance of theoretical interest for practical everyday life. However, as science has streamed into the life-world in the form of ubiquitous technology, it has brought an infinite variety of possible objects of interest into the everyday world. Pressure now arises not necessarily from our curiosity and the painful awareness that we won't be able to satisfy it within our lifespan, but also from social pressure. This phenomenon implies the pressure to keep up, the canonization of experiences, the so-called FOMO—fear of missing out, the fear of missing something.

As a result, problems are now being diagnosed in various contexts characterized by the conflict between the anthropological characteristics of limited attention on the one hand and the constant human urge to extend our range on the other. The problems are so urgent and widespread that they have found their way into societal debates. Before this, they have been present particularly in the field of sociology. A sociological milestone were Hartmut Rosa's considerations on acceleration in 2004 (2013); psychopathology around Thomas Fuchs followed up in 2018 with the thesis of the 'overburdened subject.' For many years, mindfulness and deceleration have been central keywords in popular science bestsellers. Meanwhile, not only psychologists and sociologists but also economists and computer scientists are calling for a more conscious use of digital technologies such as social media, for example with the help of a temporary 'digital detox' or long-term 'digital minimalism' (cf. Newport, 2019).

Paying attention to one thing means losing sight of another: this is the basic principle of attention (Blumenberg, 2002). Attention as a limited human resource is not only a challenge for the theoretically widely interested individual, that is a "creature with a finite lifespan" but "has infinite desires" (Blumenberg, 1986, pp. 71–72). It has become a problem for humankind in successfully designing and using technology. The reason for this is what Blumenberg calls 'working at each other's hands in a murderous way': the use of technology requires not only material resources but also the immaterial resource of attention to maintain an overview of the consequences of technology and to be able to respond to them through responsible use and design of technology.

If one now traces the creation of technologies that produce 'consciousness-like machines' and artificial worlds back to the anthropological needs formulated by Blumenberg, one can see that the latter 'work at each other's hands in a murderous way', but in a different way than Blumenberg analysed. The 'worlds on demand' in the broader sense, i.e. the digital and hybrid artificial worlds, have—among other consequences—strong material effects on the world. To counter these, we would need additional attention and prioritisation, while the artificial worlds tend to lead to mental overload. We actually need a clear mind to cope with the complexity of our technologised world. However, in the form of artificial worlds, worlds on demand, which are part of the hybridisation of our living environment, technology and the consequences of technology are now competing for our attention.

7 Conclusion: Discovering a Hidden Potential of Phenomenology

So we can conclude: unburdening the world with 'worlds on demand' did not work; even if the worlds are consumed 'at home,' they strongly influence the 'real' world. There are two ways in which the creation of virtual worlds has led to a change in the real world, and not in the sense of this desired relief: Firstly, the creation of virtual worlds is characterized by very real consequences as well as by very real demands on the real world. The technical development that enabled the establishment of virtual worlds has led to an effect contradictory to Blumenberg's vision: the world has not been unburdened from mankind, but is confronted with new burdens, e.g. a huge requirement on energy and resources in general that caused environmental side-effects. Secondly, today one can see an 'overpopulation' not only of the physical world by material objects but also of the world as a sphere of significance. Despite all benefits, the hybrid and virtual worlds we have today require mental and material resources, generate waste, and they also shape society. They must therefore be used and designed responsibly. However, this is often neglected. One reason for this may be that, if Blumenberg's thoughts are thought through further, the willingness to engage in them emerges from a deep human desire to transform reality into a deeply human-designed one through technology. It even goes beyond undesirable side-effects that can take the form of those two man-made phenomena presented in the manuscript from the estate—the production of objects and consumer behavior—'working at each other's hand in a murderous way.'

Once again, the point here is not to show whether Blumenberg was right with his prediction or whether his proposed solution would be viable in practice today (even though MacKinnon is proposing something very similar in 2021). What Blumenberg proposes in his fragment as a solution for world-clogging consumerism undeniably has a parodic touch. Rather, the aim was to show which of today's challenges were not foreseeable in this way, but which surprisingly tie in very well with observations made at that time. The idea alludes to problems of sustainability that have existed for a long time but are now more than ever in the public consciousness and the subject of social and political debate. Suddenly it is obviously important to talk about attention, and specifically about attention as a contested, limited resource. When distraction threatens vital processes, daily work and well-being, the question of sound attention management becomes existential for every individual. This need is reflected in theory, which must now explain why the smooth functioning of technology may enhance human potential in specific regards, but affects the human life-world in a way that might not be desirable for its inhabitants.

Observations of this kind have long seemed banal to us, so often have they been expressed and promoted in recent years. However, the time capsule from Blumenberg's estate sheds some new light on them because describing those contemporary phenomena with a vocabulary that leads deep into Blumenberg's thinking reveals a potential of phenomenology that it did not have during the lifetime of its founder Edmund Husserl. At that time, the phenomenological program was under great pressure to justify itself given the prospering positive disciplines and even against the background of the still comparatively diffuse crisis of European science. Today, we can observe that we are again at a historical turning point that confronts phenomenological-anthropological historiography with new challenges, but also new potentials—and that applies to Blumenberg's works in a strong sense. Blumenberg's anthropology, as it is strongly reacting to Husserl's thoughts, can help to understand the development of the digital transformation of the life-world as a historical phenomenon that emerges from processes of scientification and mechanization of the life-world on the one hand and on the other hand explain how humanity is thrown back onto challenges that have always existed but are now being exacerbated. This also means that topics previously dealt with primarily in philosophy are now becoming highly relevant to society.

In "Phenomenological Aspects on Life-World and Technization," Blumenberg says:

> Technization [...] begins [...] to regulate the life-world by making the sphere in which we do not *yet* ask questions identical with that in which we no *longer* ask questions [...]. The process of technization discloses its own 'teleology' [...] not only by producing things and achievements but also by making producible the unproducible, namely obviousness (2020, pp. 386 f.).

This tendency aims at a life-world designed through and through based on human meanings, full of comfort and assistance, security and availability, but without immediacy, without imponderables.

The tendencies which were already diagnosed by Husserl, have become more acute in this development: mathematization is progressing; psychologization, the scientific explanation of mental situations, is also continuing; but there is also a new need for phenomenological research. The work of the phenomenologist to get a theoretical grip on these supposedly self-evident facts, not only in their diversity but above all in their complexity, is given new material, new objects.

What is new, however, is that the endeavor to understand what is happening is no longer limited to philosophy. It is taking on an unprecedented social relevance: society's need to understand what is happening is growing. And this is not primarily out of theoretical curiosity, but because it has become a practical necessity to see what is increasingly sinking into obscurity, disappearing behind the processes, withdrawing into smooth, glossy casings or even into server farms and practices. We need to understand the consequences of the hybridization of the life-world, which we call digitalization, to shape it responsibly. This ranges from a careful and efficient assessment of the consequences of technology to efforts to reflect technology in more differentiated ways.

In my contribution, I have tried to show how Blumenberg's thinking, with its sharp eye for supposed and artificial obviousness, can help us today to lift both the technologies and our practices, in which they are embedded to the point of invisibility, out of questionlessness; namely by helping us to understand how, on the one hand, anthropological needs and peculiarities have led to technization of the life-world and cling to it and, on the other hand, how this technization can come into conflict with precisely these needs due to precisely these peculiarities. While practical approaches such as technology ethics, technology assessment or governance and regulation deal with the question of how negative technological consequences and risks can be managed and how technologies can be designed and used responsibly, a continuation of the considerations presented here can help us to understand this ambivalence of the technization of the life-world and perhaps even to develop insights that can contribute to a more responsible approach to it.

References

Blumenberg, H. (1979). The concept of reality and the possibility of the novel. In R. E. Amacher & V. Lange (Eds.), *New perspectives in German literary criticism: A collection of essays* (pp. 29–48). Princeton University Press.
Blumenberg, H. (1986). *Lebenszeit und Weltzeit*. Suhrkamp.
Blumenberg, H. (2002). *Zu den Sachen und zurück*. Suhrkamp.
Blumenberg, H. (2015). *Schriften zur Technik*. Suhrkamp.
Blumenberg, H. (2020). Phenomenological aspects on life-world and technization. In H. Bajohr, F. Fuchs, & J. P. Kroll (Eds.), *History, metaphors, fables. A Hans Blumenberg reader* (pp. 358–299). Cornell University Press.
Carroll, M. (1980). *Sceptical sociology*. Routledge & Kegan Paul.
Eisler, R. (1904). *Wörterbuch der philosophischen Begriffe* (Vol. 1). Mittler.
Feger, S. (2020). Hans Blumenbergs Wirklichkeitsbegriff aus phänomenologischer Perspektive. *Phänomenologische Forschungen, 1*, 41–64. https://doi.org/10.28937/1000108511
Fuchs, T. (2018). *Ecology of the brain. The phenomenology and biology of the embodied mind*. Oxford University Press.
Günther, G. (1976) *Beiträge zur Grundlegung einer operationsfähigen Dialektik. Erster Band: Metakritik der Logik – nicht-aristotelische Logik – Reflexion – Stellenwerttheorie – Dialektik - Cybernetic ontology – Morphogrammatik – transklassische Maschinentheorie*. Meiner.
Klein, A. (2017). *Zwischen Grenzbegriff und absoluter Metapher: Hans Blumenbergs Absolutismus der Wirklichkeit*. Ergon.
MacKinnon, J. B. (2021) *The day the world stops shopping. How ending consumerism gives us a better life and a greener world*. The Boadly Head.
Nassehi, A. (2024). *Patterns: Theory of the digital society*. Polity Press.
Newport, C. (2019) *Digital minimalism. Choosing a focused life in a noisy world*. Penguin Business.
Niehues-Pröbsting, H. (2022). Annäherung an Blumenbergs Philosophieverständnis. *Deutsche Zeitschrift für Philosophie, 70*(1), 64–86. https://doi.org/10.1515/dzph-2022-0003
Philippi, M. (2010). Ein Spiel mit Selbstverständlichkeit(en). Formal-inhaltliche Übergänge in Blumenbergs philosophischen Miniaturen. In A. Fragio Gistau & D. Giordano (Eds.), *Hans Blumenberg. Nuovi Paradigmi d'analisi* (pp. 237–261). Aracne.
Quent, M., & Plattner, G. (2024). How can we shape immersive digital worlds democratically? *TATuP – Zeitschrift für Technikfolgenabschätzung in Theorie Und Praxis, 33*(2), 62–65. https://doi.org/10.14512/tatup.33.2.62
Rosa, H. (2013). *Social acceleration: A new theory of modernity*. Columbia University Press.
Weiser, M. (1991). The computer for the 21st century. *Scientific American Magazine, 265*(3), 94–104.

Martina Philippi did her Ph.D. in philosophy on Edmund Husserl's phenomenology. She worked at Leipzig University and Ruhr-University Bochum and is currently a postdoc researcher in the Collaborate Research Center "Constructing Explainability" (SFB/TRR 318) at Paderborn University. Her main topics are phenomenology and ethics of Artificial Intelligence and Digital Transformation. Book publications: *Metaphorologie, Anthropologie, Phänomenologie. Neue Forschungen zum Nachlass Hans Blumenberg* (together with Alberto Fragio and Josefa Ros Velasco, Alber Verlag, Freiburg/Br., 2020); *Selbstverständlichkeit und Problematisierung. Husserls Programm der Phänomenologie* (Fink Verlag, Paderborn, 2023).

Correction to: Hans Blumenberg's History and Philosophy of Science

Josefa Ros Velasco

Correction to:

A. Fragio et al. (eds.), Hans Blumenberg's History and Philosophy of Science, Studies in History and Philosophy of Science 63, https://doi.org/10.1007/978-3-031-78251-0

The original version of the book was inadvertently published with incorrect affiliation of the co-editor and author of Chapter 6, Josefa Ros Velasco. This has been updated and corrected in Chapter 6.

The updated version of this book can be found at
https://doi.org/10.1007/978-3-031-78251-0

J. Ros Velasco (✉)
Department of Humanities, ESIC University, Madrid, Spain
e-mail: josros@ucm.es

A. Fragio et al. (eds.), *Hans Blumenberg's History and Philosophy of Science*, Studies in History and Philosophy of Science 63,
https://doi.org/10.1007/978-3-031-78251-0_10

The manufacturer's authorised representative in the EU is Springer Nature Customer Service Centre GmbH, Europaplatz 3, 69115 Heidelberg, Germany. If you have any concerns regarding our products, please contact ProductSafety@springernature.com

Printed and bound by CPI Group (UK) Ltd, Croydon, CR0 4YY
30/09/2026
02225513-0018